T0069284

MAP OF THE MOON

300 INCH

MAP OF THE MOON

DRAWN BY

H. PERCY WILKINS. F.R.A.S.

Director of the Lunar Section of the British Astronomical Association from 1946.

Hon. President of the Sociedad Astronómica de España y América, President Cardiff & District Astronomical Society, Fellow British Interplanetary Society, Memb. American Assn. Lunar & Planetary Observers, &c.

THIRD EDITION – 1951.

100 in. REPRODUCTION IN 25 SECTIONS.

The details shown on this map have been taken from the finest photographs, drawings by the most eminent observers and as the result of personal observations, commencing in 1909, with telescopes of various apertures, including a first-class reflector of 15¼ inches aperture. Based on the measures of Saunders & Franz, supplemented by those of the Author. To the advancement of Selenography and to the immortal memory of Mayer, Schröeter, Lohrmann, Madler, Schmidt, Neison, Elger, Goodacre, Pickering and all the selenographers who by their labours in the past have advanced our Knowledge, this work is dedicated by the Author.

Orthographic Projection at Mean Libration. Scale 21·6 miles per inch. Two Libratory Sections.

KEY:– ◎ Craterlet, ⊙ Craterpit, ⊕ Cratercone, o Hillock, ═ Clefts, ══ Ridges, ······ Light Streaks, ⋕ Dark Variable Spots. Sections numbered as on Index Map below.

S

25 24 23 22 21

10 9 8 7 20

W 11 2 1 6 19 E

12 3 4 5 18

13 14 15 16 17

N

PUBLISHED BY THE AUTHOR AT 35, FAIRLAWN AVENUE, BEXLEYHEATH, KENT, ENGLAND.

Author:– Born 1896, Dec. 4, at Carmarthen, S. Wales, Gt. Britain, issue of William John Wilkins (died 1932, July 30, date of commencement of Map) & Alice Wilkins (née Richens, died 1951, March 30, date of completion of Map, 3rd Edition revision). Index Sheet compiled by Mr. P. A. Moore. F.R.A.S.

Latest impression includes detail discovered April 1952 with the 33 in. Meudon Refractor.

made by fitting lenses into cardboard tubes; later, he would learn to grind mirrors to build more serious telescopes. After serving in the First World War (when he insisted on taking a small telescope with him), he became an engineer, eventually working at the Ministry of Supply, which coordinated the supply of equipment to the British armed forces.

Outside of work he became an astronomer with an international reputation. He was given honours including the Spanish Civil Order of Alfonso X, awarded for distinguished achievement in arts and sciences. From 1946 to 1956, he was Director of the Lunar Section of the British Astronomical Association (BAA), which has sought to support amateur astronomers in making scientifically valuable observations since 1890. Amateur astronomy could be a very serious business. For people like Wilkins, it involved fierce commitments of time and energy to nights observing, to telescope construction, to publication and to correspondence. Wilkins's own work was based on existing lunar maps and photographs, drawings from a network of observers across the world, and his own observations, made both at home and at major observatories.

It was principally from his garden in Bexleyheath, in the suburbs of London, that Wilkins checked the detail of different features which he had examined in photographs, before they were included in his map. He sketched the way that shadows fell on the lunar surface, in order, eventually, to help interpret the shapes of mountains, valleys, craters and domes. To observe the Moon was to observe the way that changing light and shadow revealed and hid again its features. As Wilkins commented, 'much of the finest detail, the most difficult to depict, but by far the most fascinating, vanishes before one's eyes, dying, so to speak, even while it is born.'* A selection of Wilkins's detailed sketches are also included in this volume.

His lunar work was not without controversy, and his reputation suffered when in 1953 he publicly confirmed the existence of a natural bridge on the Moon, first observed by US astronomer John J. O'Neill. This was in fact an illusion, caused by the way shadows fell in a particular region, giving the appearance of light shining under a natural arch. Worse, Wilkins was quoted in a way which made it sound like he suggested the bridge could have been constructed by intelligent life. Criticised

* Hugh Percy Wilkins, 'Notes on Lunar Drawing', The Moon 3:2 (1954):43

by astronomical colleagues, he resigned as director of the BAA Lunar Section. Scrutiny of his cartographic work has also not always been positive, with his work sometimes criticised for being overcrowded and difficult to read. Wilkins worked in a tradition which valued moon maps of ever greater detail at a moment when, with early robotic and crewed missions into space, the purpose of pursuing depictions of finer and finer topographic detail through larger and larger telescopes was called into question. However, Wilkins's map remained the most detailed of its time, and it was used to help relate the first photographs ever taken of the far side of the Moon by the Soviet spacecraft Luna 3, to the nearside, and was purchased by NASA during the Apollo programme. The amateur astronomical journal *The Strolling Astronomer* insisted of the map that 'all students of the Moon should have it', and even published a reduced version of the map for people to refer to at the telescope.*

Wilkins died in 1960, shortly after retiring, and nine years before people first set foot on the Moon. By the end of his life, the politics of the Space Age was driving new ways of examining the lunar surface. Wilkins always insisted, however, that whatever the possibilities were, 'the best way to see the Moon is through a telescope'.†

The National Maritime Museum acquired this map in 2006, in addition to a selection of Wilkins's notebooks, letters and photographs, which are now part of the Museum's collection. Wilkins's work, with its distinctive graphic style, is a testament to his commitment to astronomy and to the importance of the work of amateurs in this field. It is also an invitation to us all to explore the geography of our nearest celestial neighbour, reproduced here in fascinating detail.

---

* 'Do you want a Map of the Moon', *The Strolling Astronomer*, April 1950

† 'Percy Maps the Moon for Space Men', *Empire News*, 4 January 1953

INDEX MAP

The illustrations on the following pages originate from a 100-inch reproduction in 25 sections of Hugh Percy Wilkins's *Map of the Moon*. For the purpose of this publication, each section has been divided into quarters. To aid the reader, each quarter section and corresponding page number is detailed here.

Most astronomical telescopes present a south-up view, as seen below.

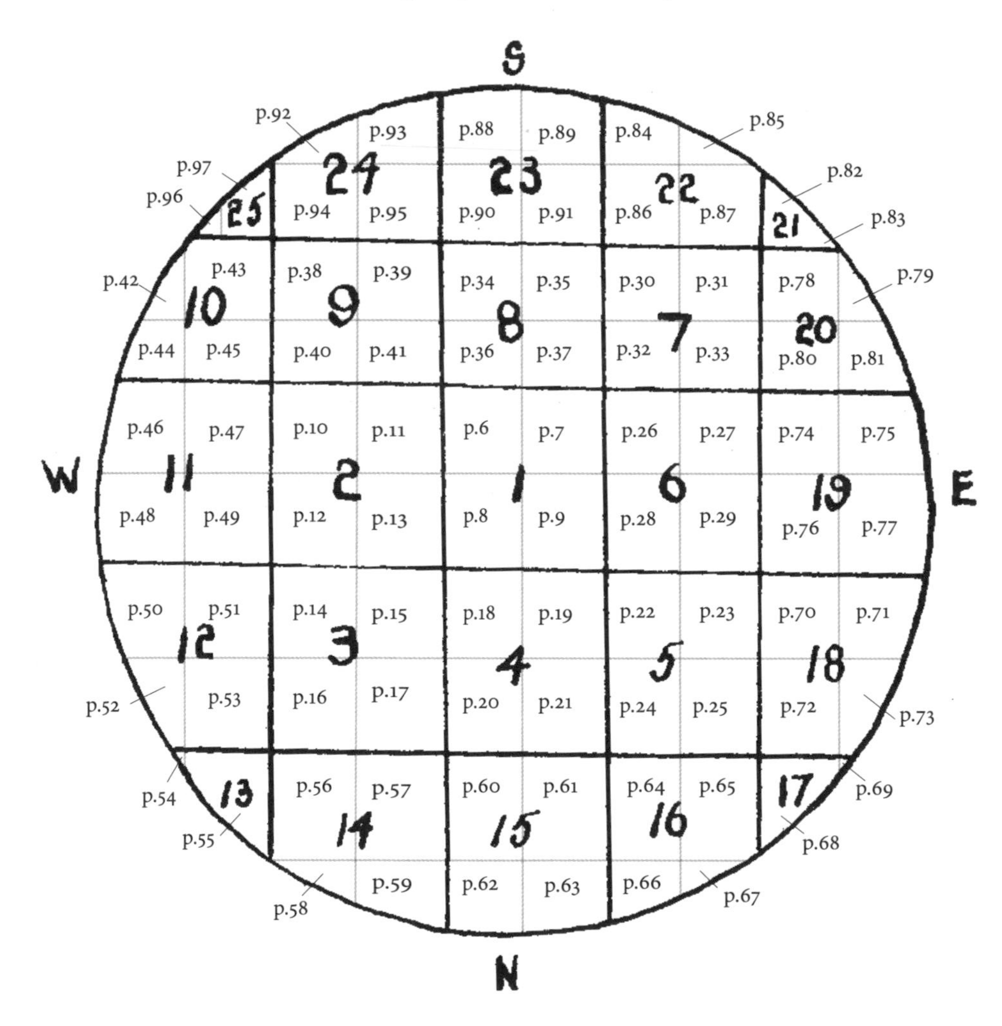

ON THE MOON

Craters on the Moon have mainly been formed by asteroid impacts. Billions of years of these impacts have also created the rocks, boulders and dust which cover the lunar surface. Lunar 'seas', or *maria* (their Latin name), are in fact plains of solidified lava. These *mare* are the dark patches which you can see on the Moon from Earth. Astronomers in Europe had been giving names to the different features they observed through telescopes since the 17th century. Gradually, an agreed system of naming emerged. From the 20th century, names have been authorised by the International Astronomical Union. By convention, plains, described as seas (*maria*), are named after weather or emotion, such as Mare Imbrium (the 'Sea of Rains') and Mare Tranquillitatis (the 'Sea of Tranquility'). Craters are named after people, generally notable scientists, philosophers, astronomers and mathematicians, and mountains are named either after people or mountains on Earth. Subsidiary features, such as smaller craters, are given the name of a larger crater nearby along with a letter, to help astronomers distinguish them. In his map, Wilkins included various unofficial names, as well as the standard ones. You can find explanations of different features mapped in each section throughout the book.

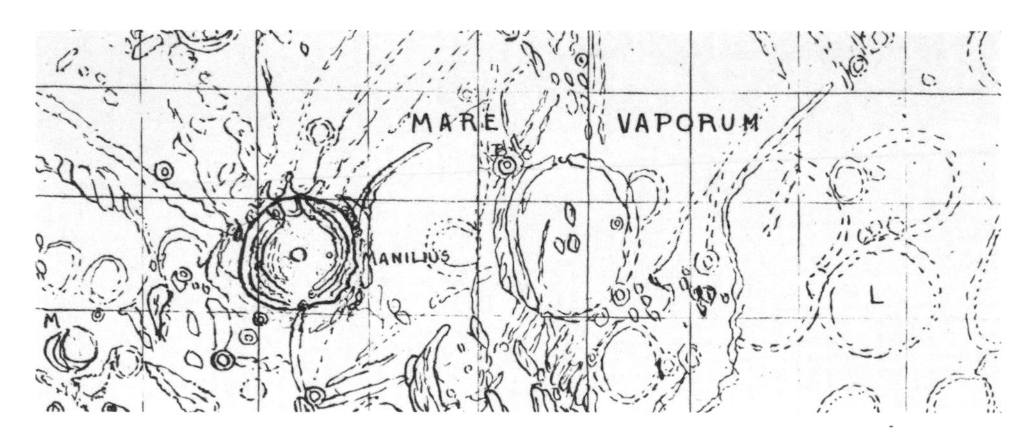

Key:- ◎ Craterlet, ⊙ Craterpit, ⊕ Cratercone, o Hillock, —η— Clefts, ══ Ridges, ······ Light Streaks, ## Dark Variable Spots.

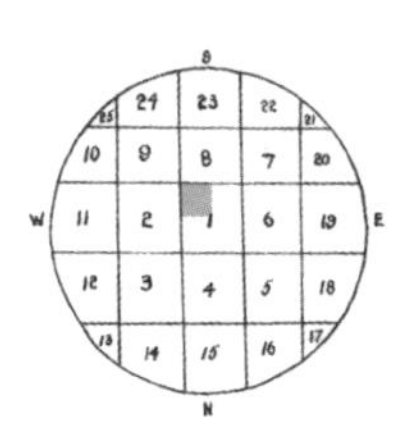

SECTION:
I:I

Key:- ⊚ Craterlet, ⊙ Craterpit, ⊕ Cratercone, o Hillock, Clefts, Ridges, ... Light Streaks, Dark Variable Spots.

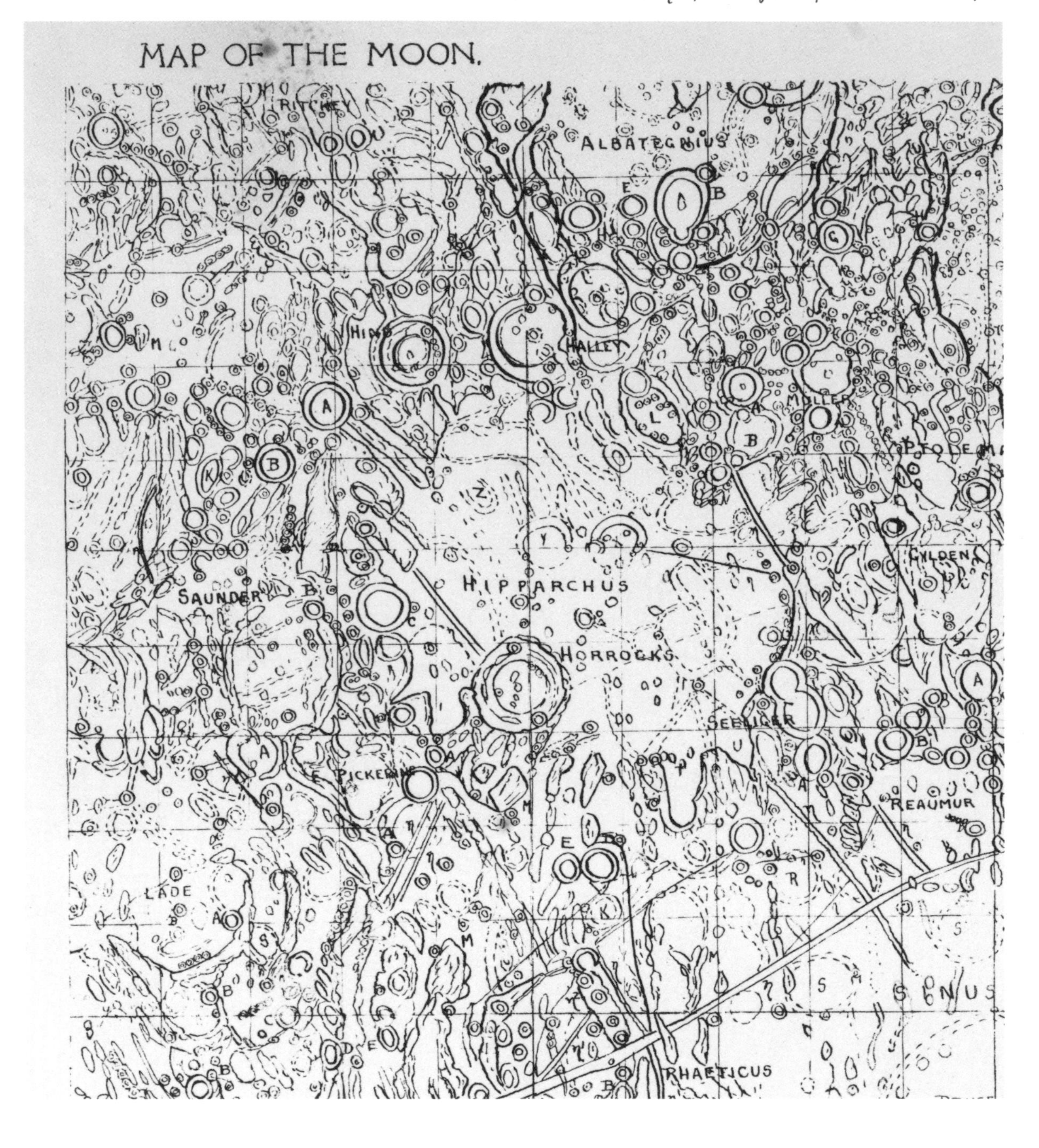

KEY:– ⊚ Craterlet, ⊙ Craterpit, ⊕ Cratercone, ○ Hillock, ─η─ Clefts, ═══ Ridges, ······ Light Streaks, ▦ Dark Variable Spots.

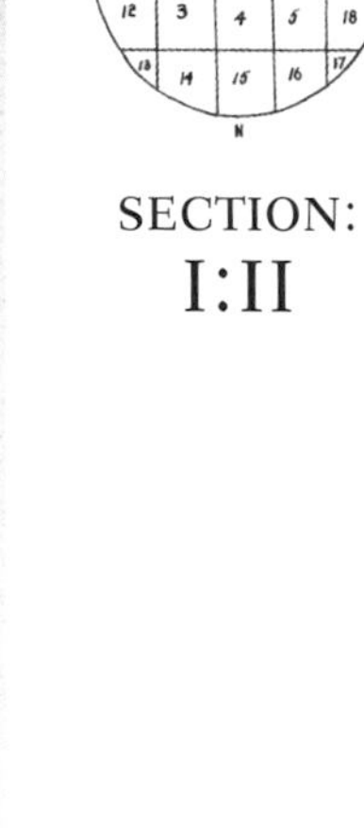

SECTION:
I:II

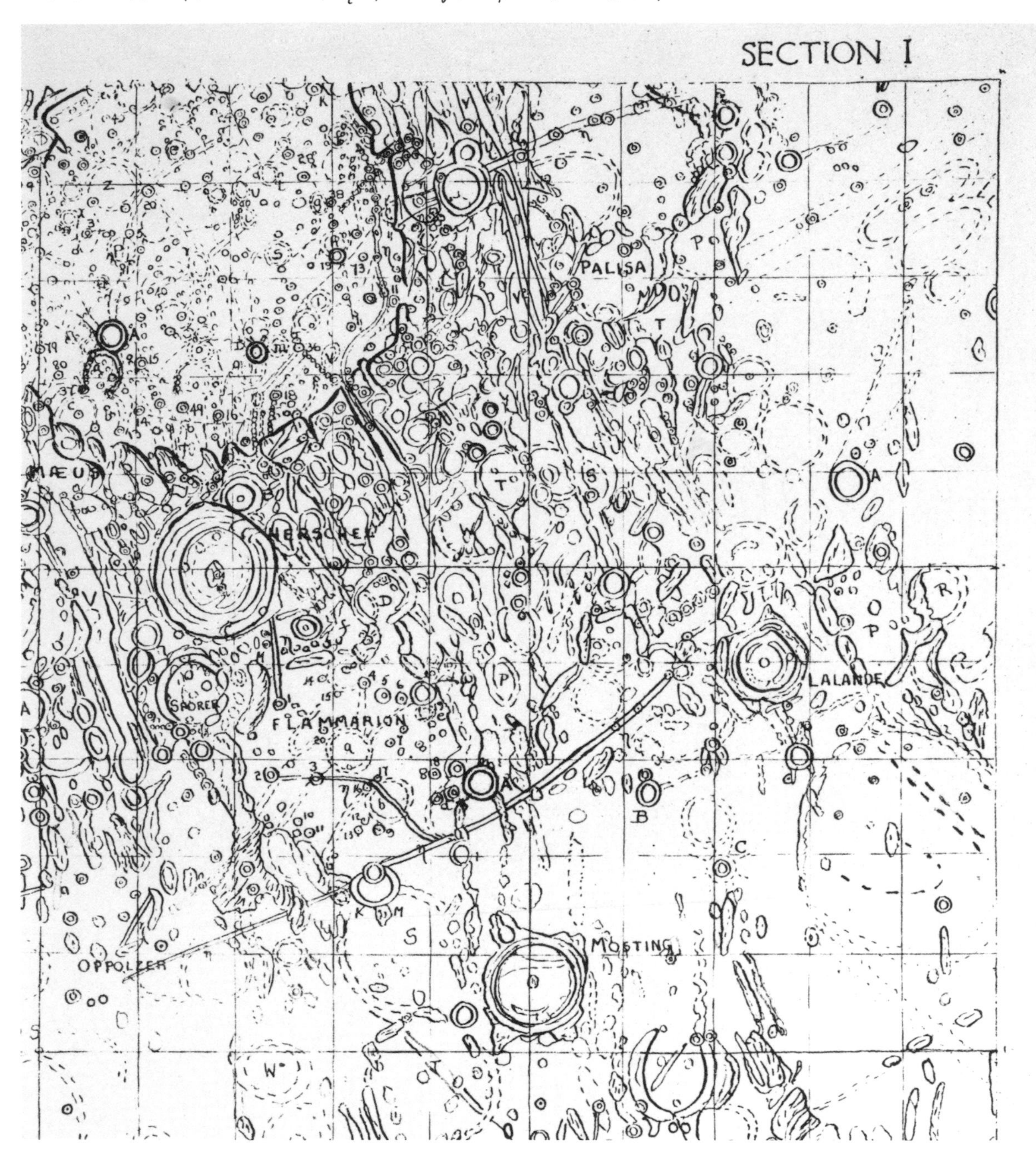

The Apollo 14 landing site (5 February 1971) was near the crater **Lalande**.

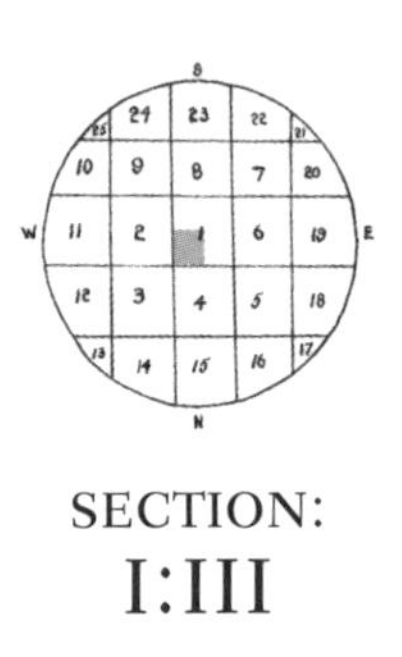

SECTION:
I:III

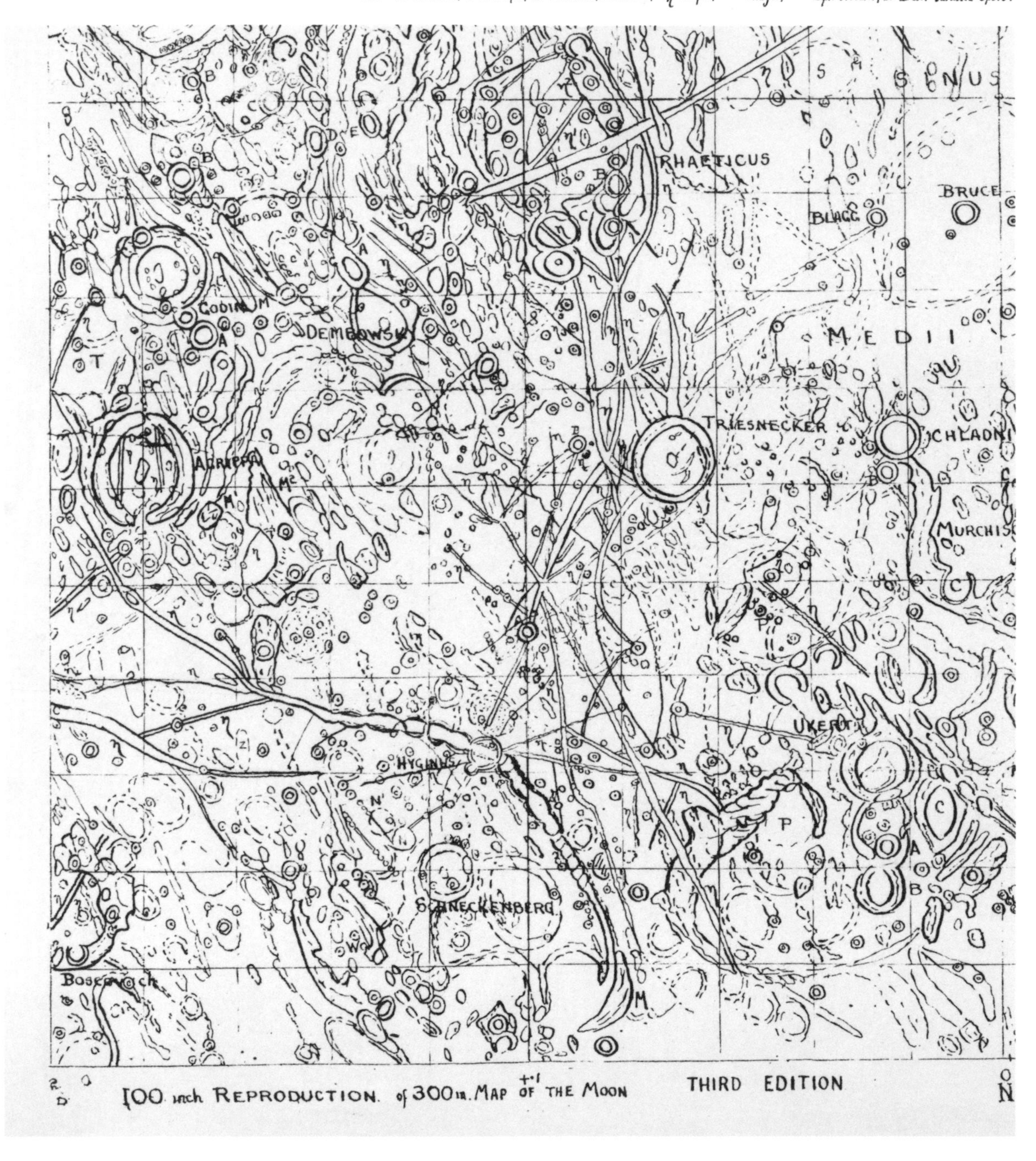

Mary Blagg (1858–1944) was one of the first five women to be elected Fellow of the Royal Astronomical Society, in 1916. Her work standardising lunar nomenclature was adopted by the International Astronomical Union in 1935.

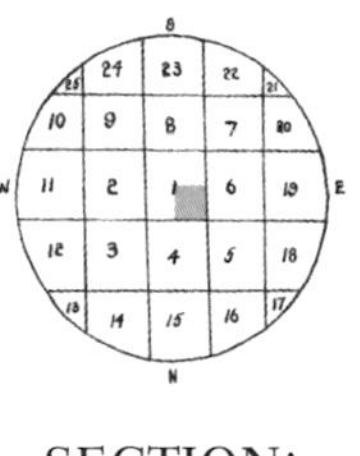

SECTION:
I:IV

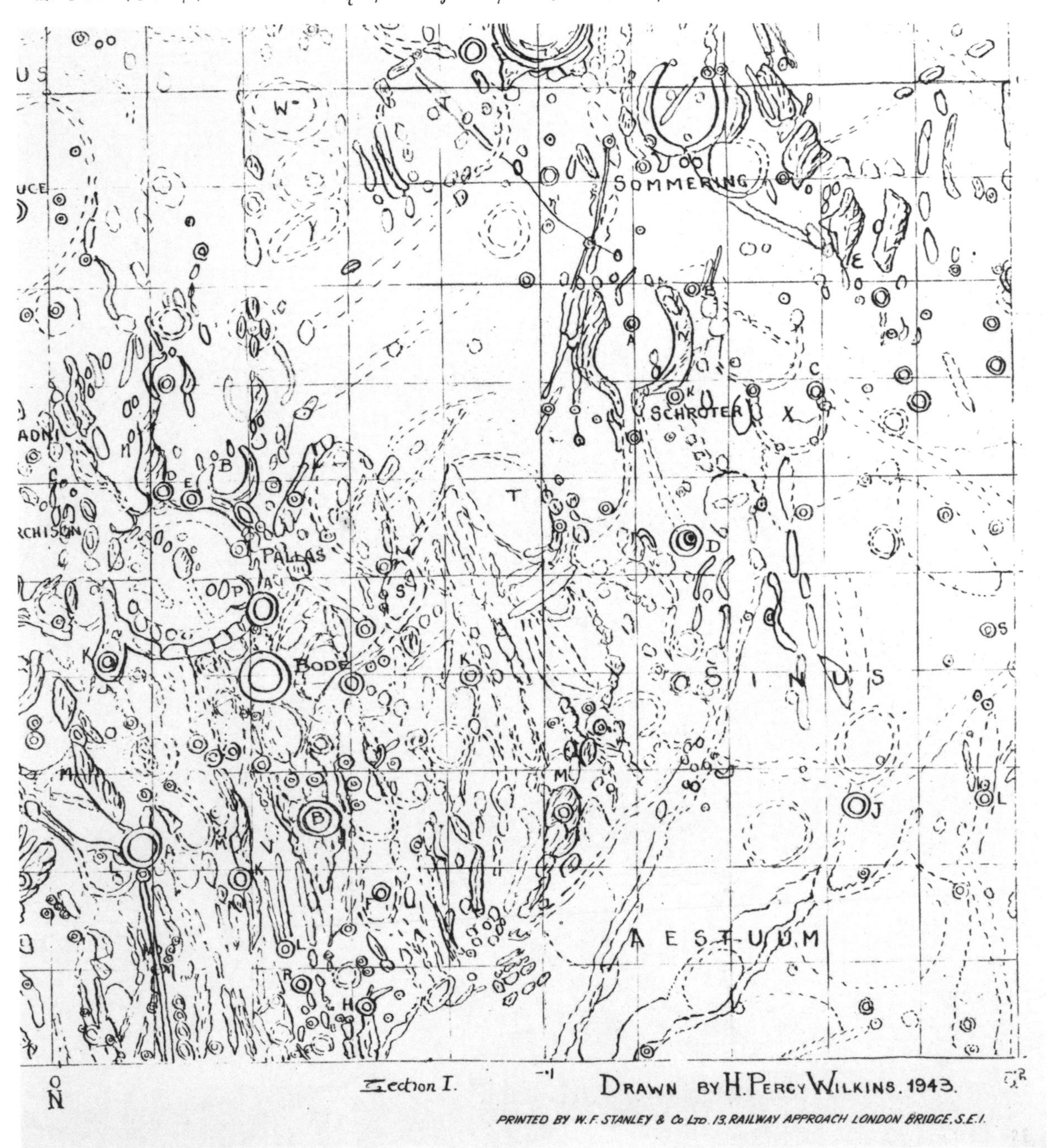
KEY:- ⊚ Craterlet, ⊙ Craterpit, ⊕ Cratercone, o Hillock, Clefts, ═ Ridges, ······ Light Streaks, Dark Variable Spots.
SOMMERING
SCHROTER
PALLAS
BODE
S I N U S
A E S T U U M
Section I.
DRAWN BY H. PERCY WILKINS. 1943.
PRINTED BY W. F. STANLEY & Co LTD. 13, RAILWAY APPROACH LONDON BRIDGE, S.E.I.

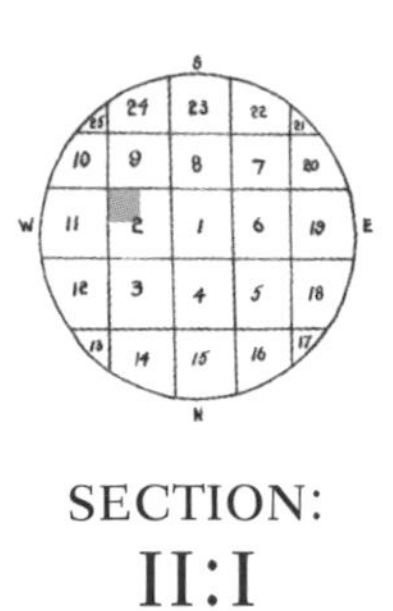

SECTION:
II:I

KEY:– ⊚ Craterlet, ⊙ Craterpit, ⊕ Cratercone, o Hillock, Clefts, ═ Ridges, ······ Light Streaks, Dark Variable Spots.

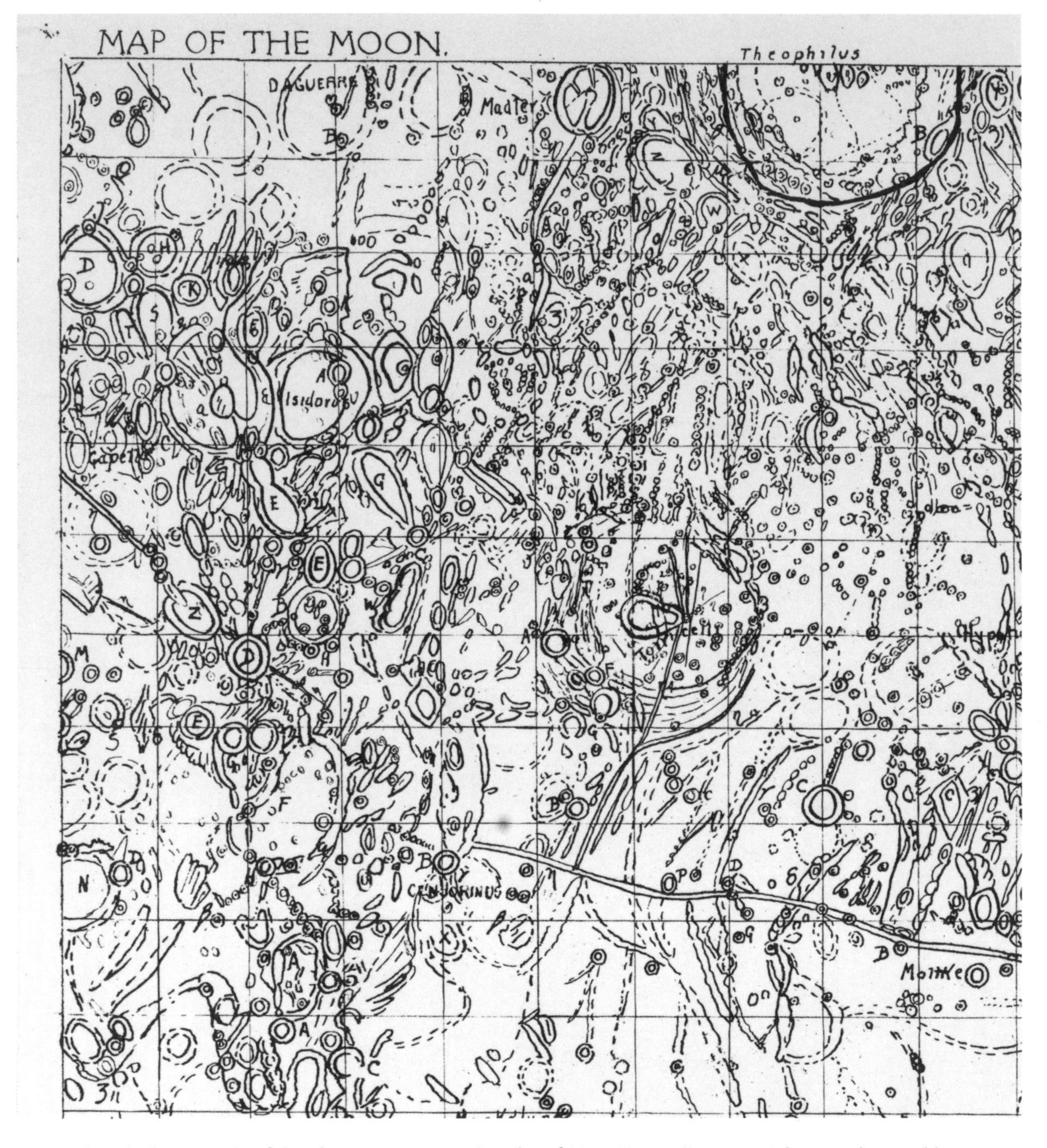

Depicted on the bottom right of this sheet is an area at the edge of Mare Tranquillitatis. In July 1969, this would become the Apollo 11 landing site, the place where a person first set foot on the Moon.

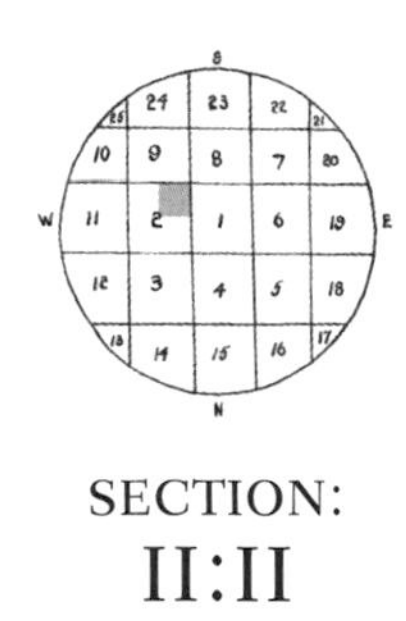

SECTION:
II:II

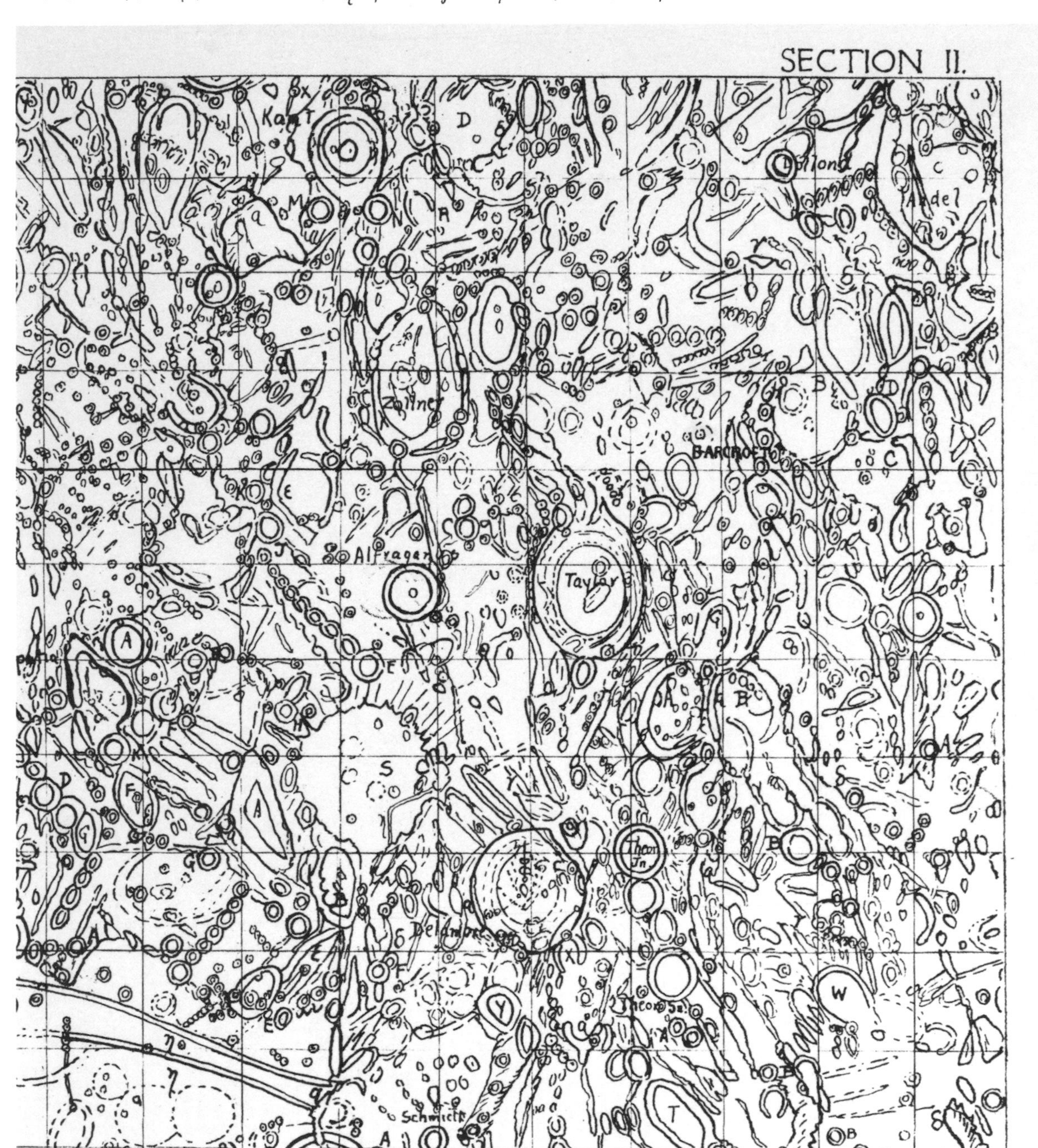

Hypatia was a 4th century CE scholar who was renowned for her mathematical and philosophical learning. She was murdered in 415 CE in the turbulent politics of her native Alexandria. The Apollo 16 landing site (21 April 1972) was located near the crater **Kant**.

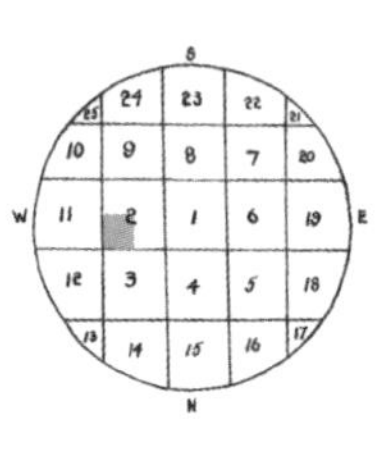

SECTION:
II:III

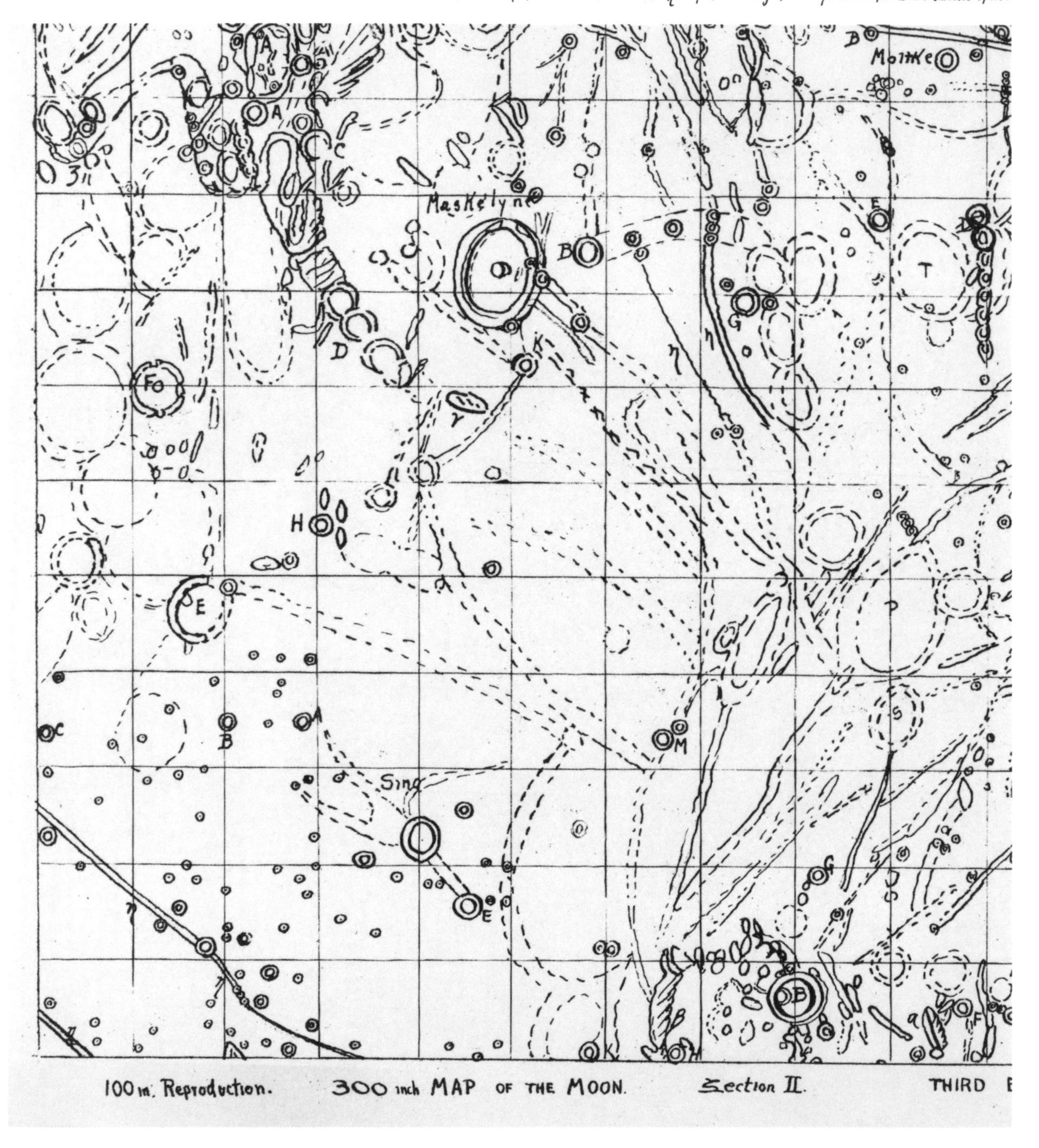
KEY:- ◎ Craterlet, ⊙ Craterpit, ⊕ Cratercone, o Hillock, Clefts, Ridges, ... Light Streaks, Dark Variable Spots.
Maskelyne
Moltke
Sin
100 in. Reproduction.
300 inch MAP OF THE MOON.
Section II.
THIRD

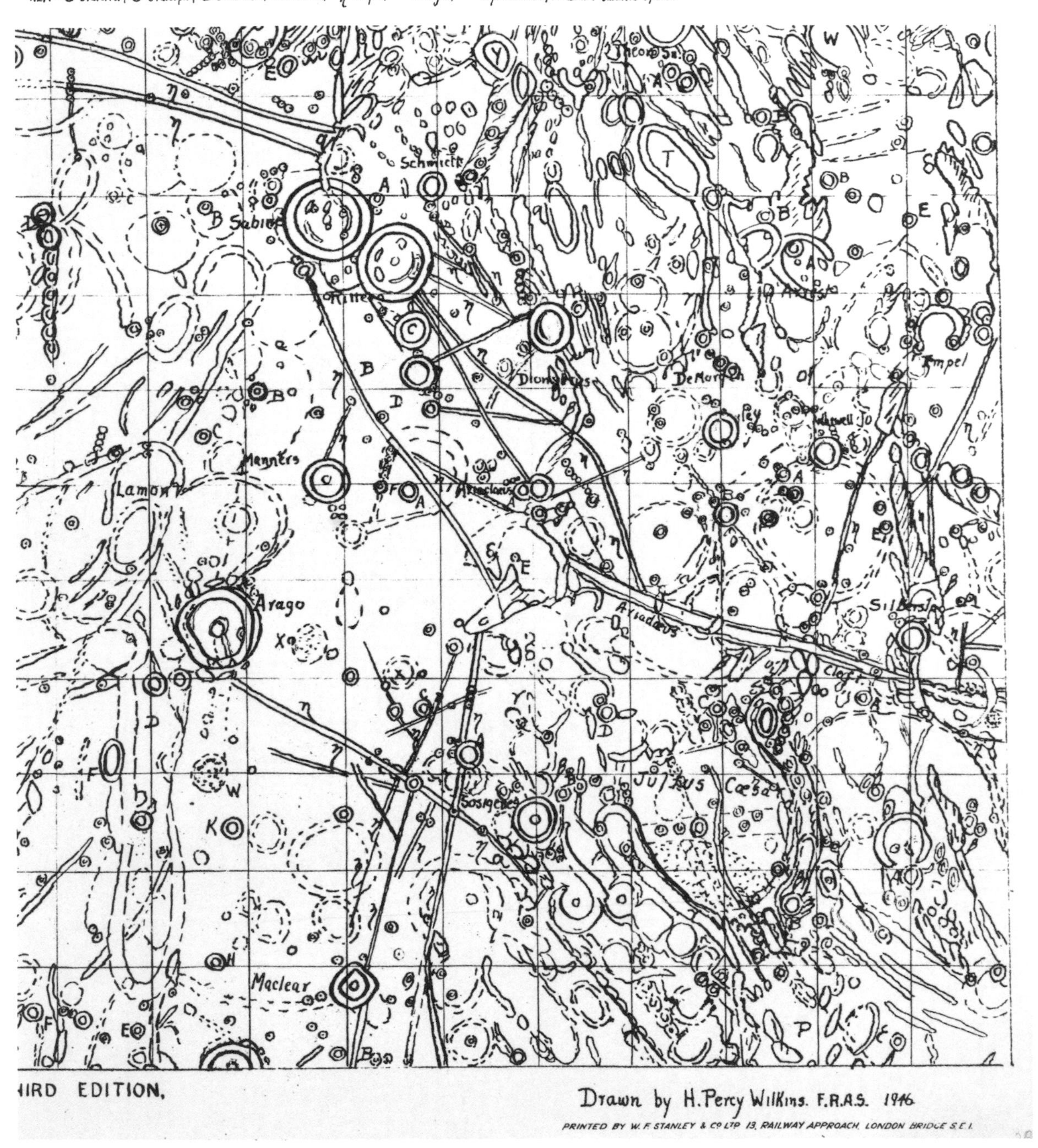
KEY:- Craterlet, Craterpit, Cratercone, Hillock, Clefts, Ridges, Light Streaks, Dark Variable Spots.
Schmidt
Sabine
Ritter
Dionysius
DeMorgan
Manners
Lamont
Arago
Arsaclus
Silberschlag
Maclear
Sosigenes
Julius Caesar
Cleft
THIRD EDITION.
Drawn by H. Percy Wilkins. F.R.A.S. 1946.
PRINTED BY W. F. STANLEY & Co LTD 13, RAILWAY APPROACH, LONDON BRIDGE S.E.1.

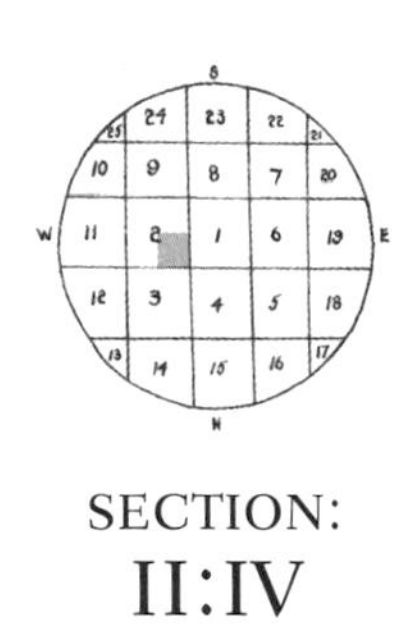
W
E

SECTION:
II:IV

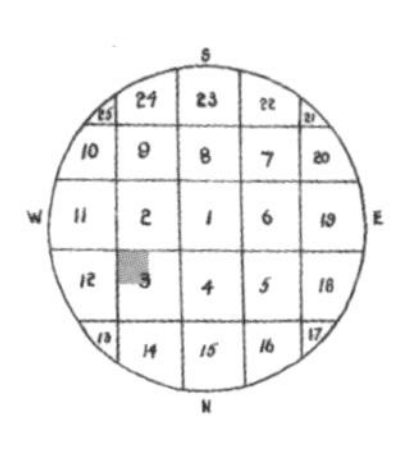

SECTION:
III:I

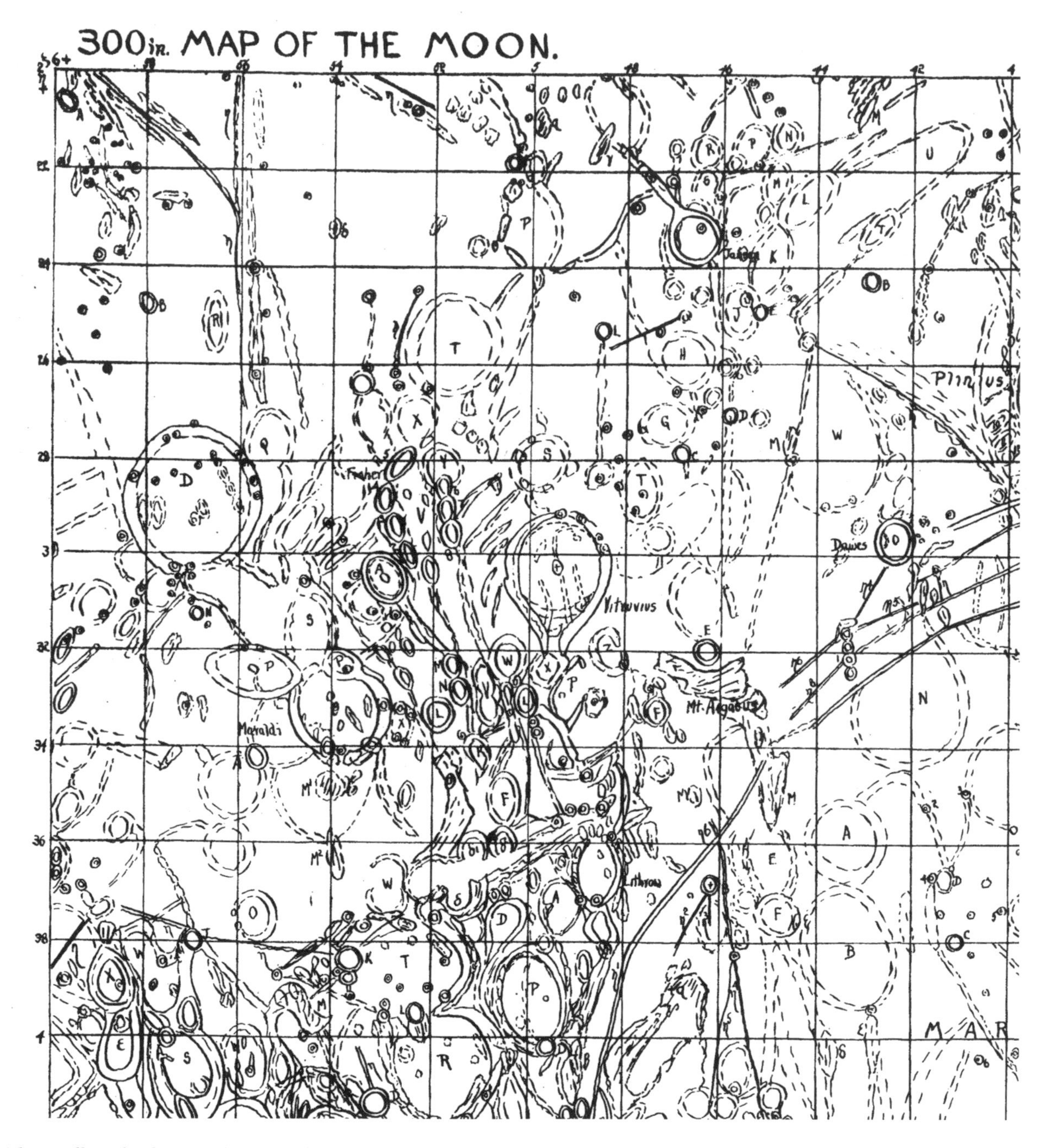

The Apollo 17 landing site (11 December 1972) was located between craters **Vitruvius** and **Littrow**.

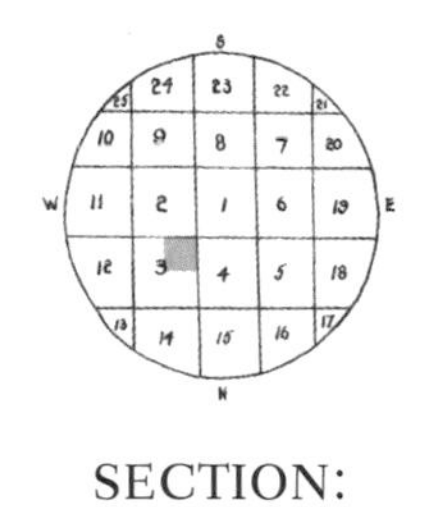

SECTION:
III:II

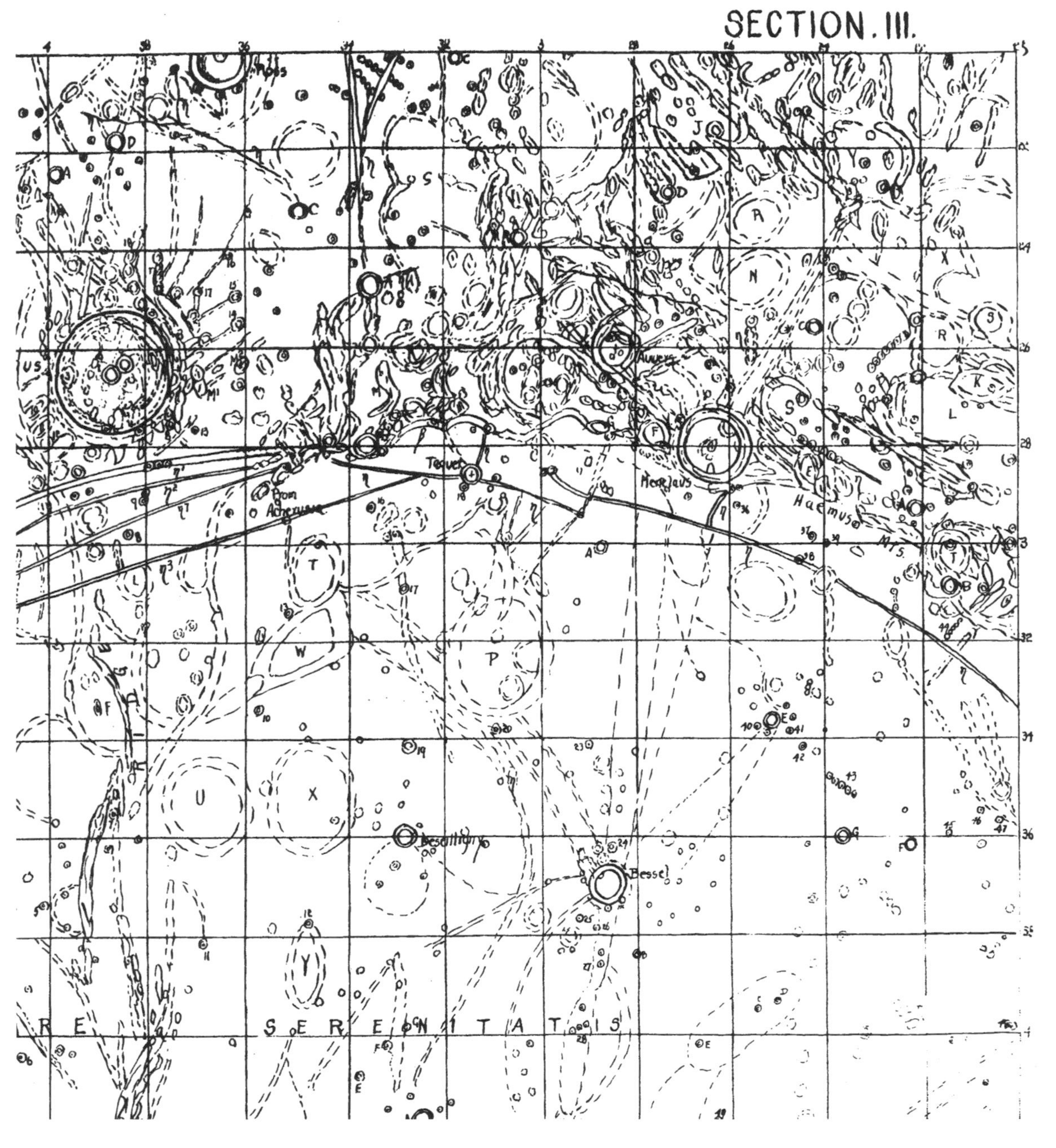

The formation of **Mare Serenitatis**, the ‘Sea of Serenity’, would have been far from serene. It is a vast plain caused by volcanic action billions of years ago, when lava spilled out from cracks in the Moon’s surface caused by asteroid impacts. Later asteroid impacts created craters such as **Bessel** and **Deseilligny.**

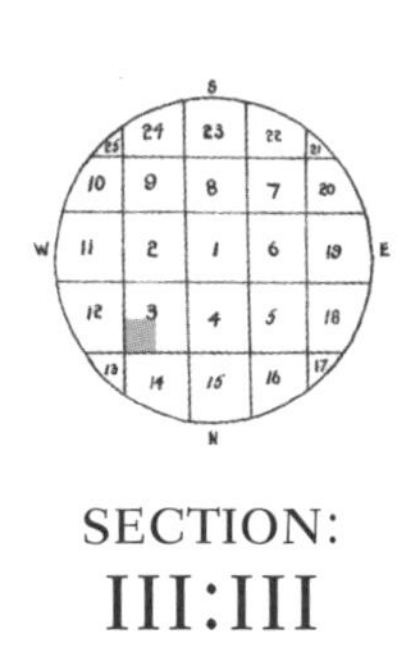

SECTION:
III:III

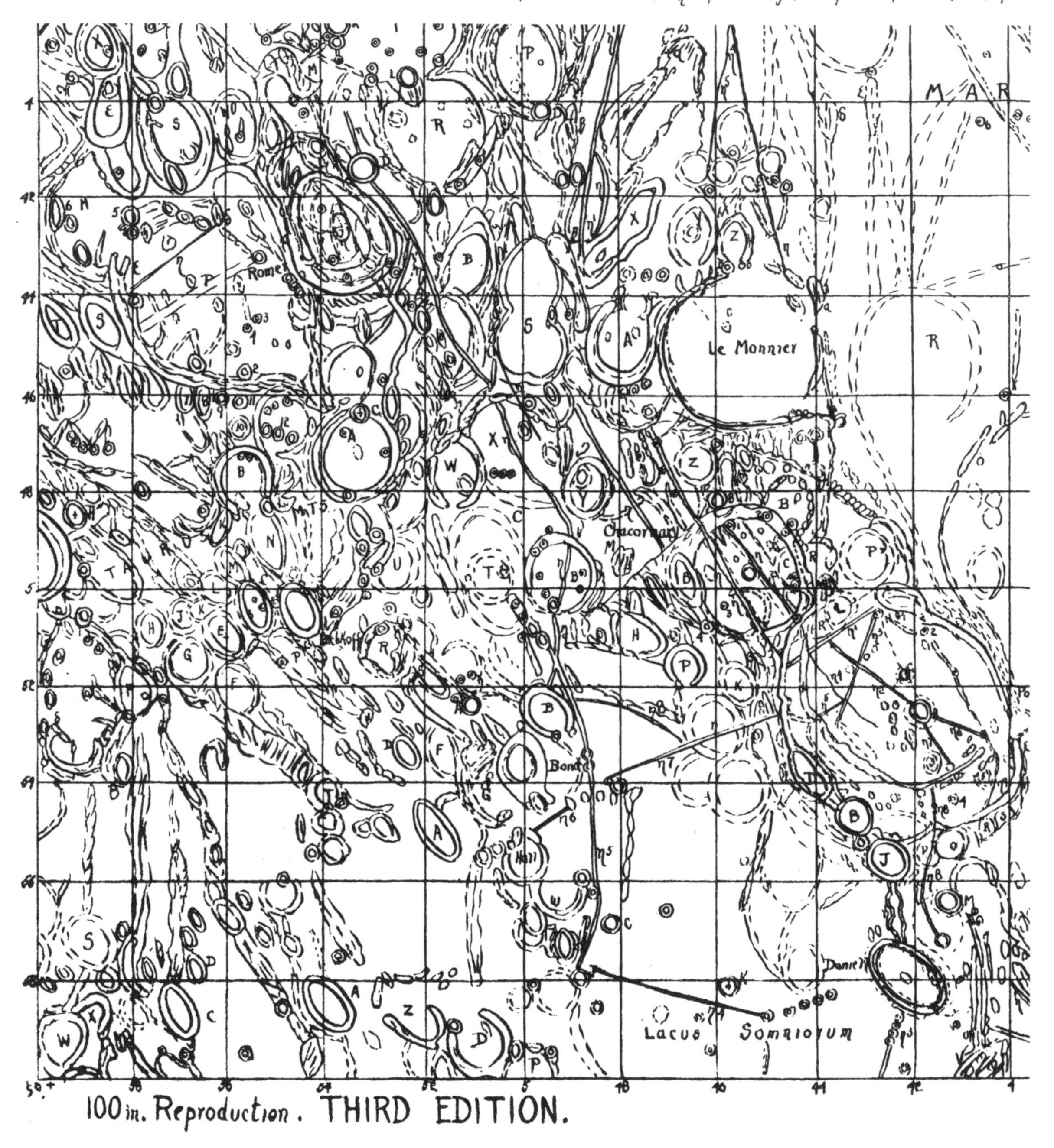

KEY:- Craterlet, Craterpit, Cratercone, Hillock, Clefts, Ridges, Light Streaks, Dark Variable Spots.
Le Monnier
Chacornac
Römer
Bond
Hall
Daniell
Lacus Somniorum
100 in. Reproduction. THIRD EDITION.

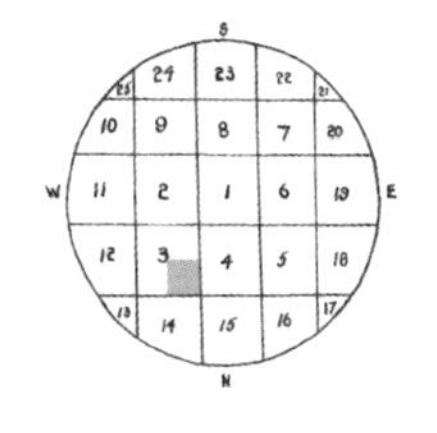

SECTION:
III:IV

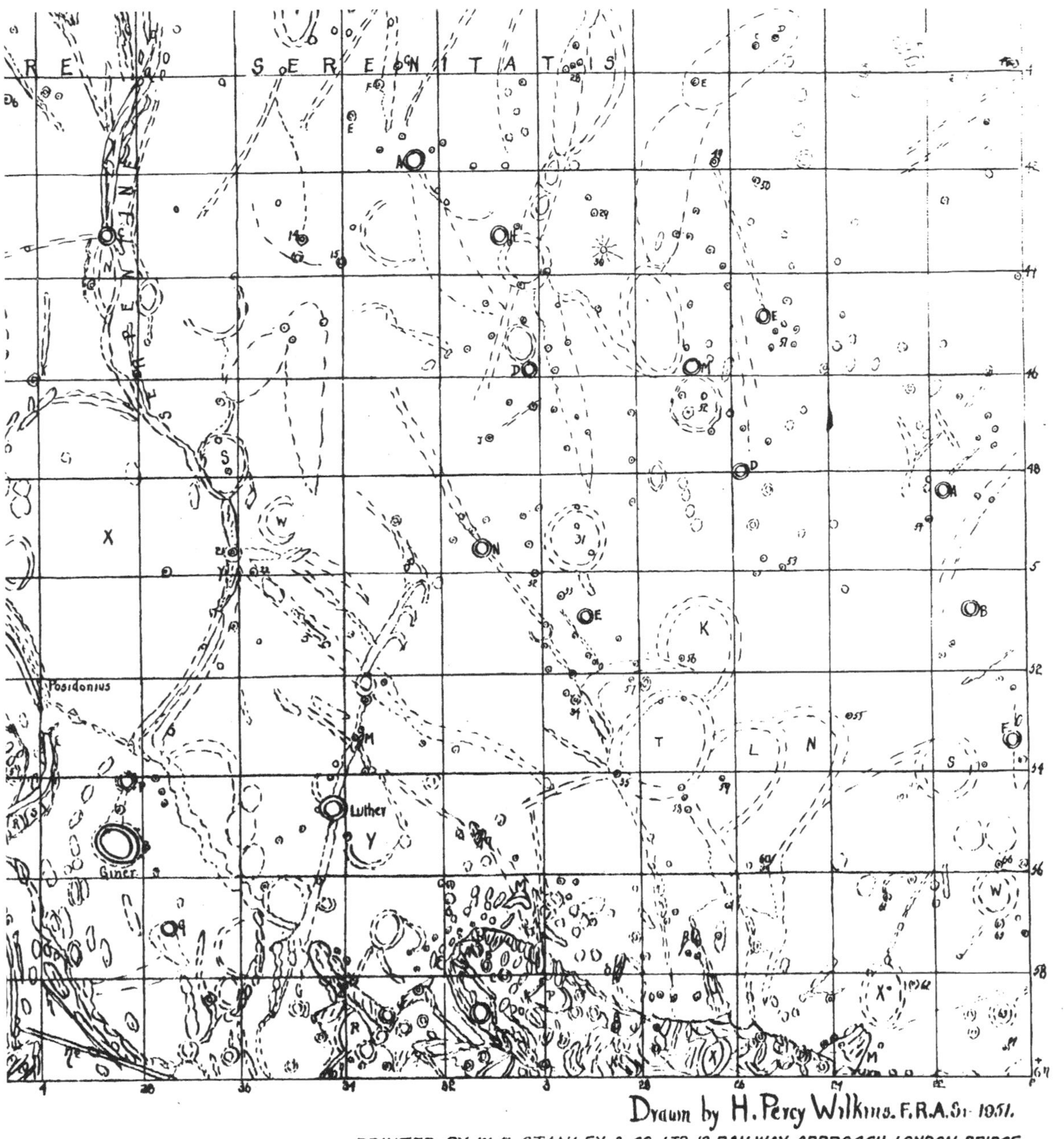

KEY:- Craterlet, Craterpit, Cratercone, Hillock, Clefts, Ridges, Light Streaks, Dark Variable Spots.
S E R E N I T A T I S
Posidonius
Luther
Giner
Drawn by H. Percy Wilkins. F.R.A.S. 1951.
PRINTED BY W. F. STANLEY & Co. LTD 13 RAILWAY APPROACH, LONDON BRIDGE. S.E.1.

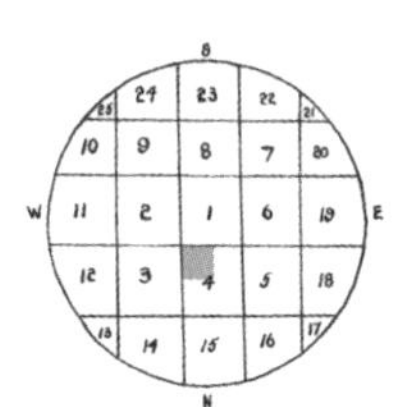

SECTION:
IV:I

KEY:- ◎ Craterlet, ⊙ Craterpit, ⊕ Cratercone, o Hillock, Clefts, ══ Ridges, ······ Light Streaks, Dark Variable Spots.

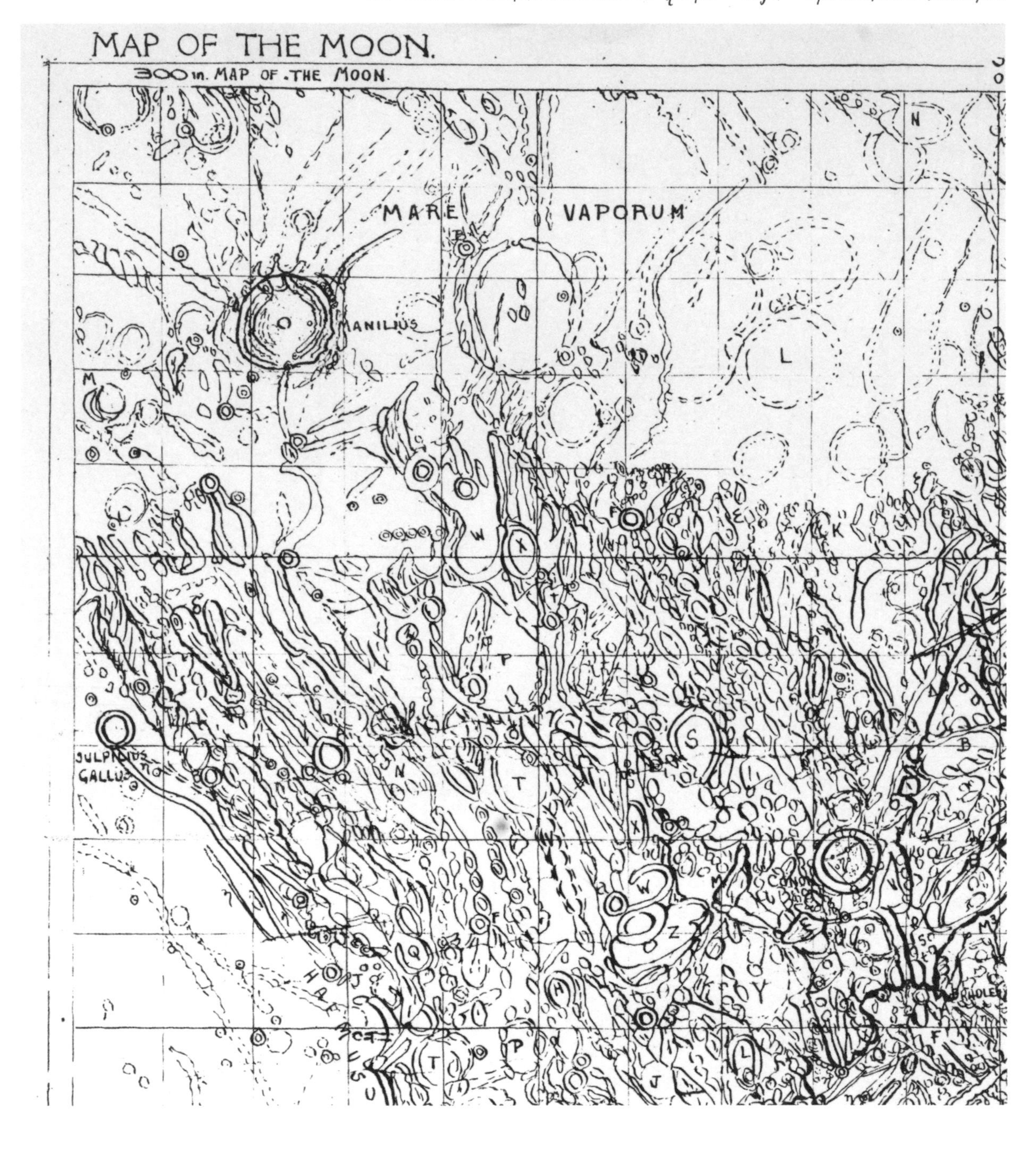

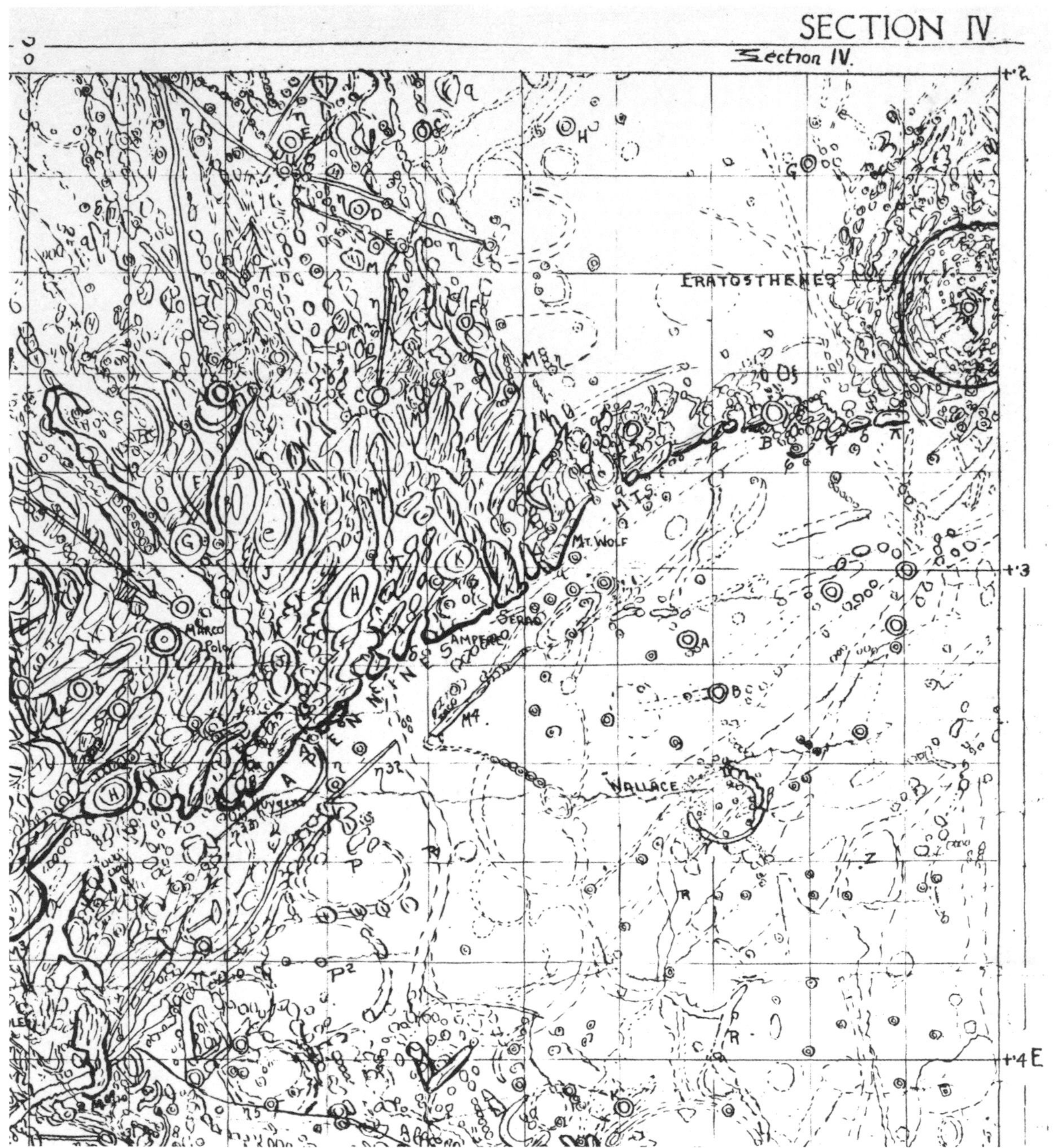

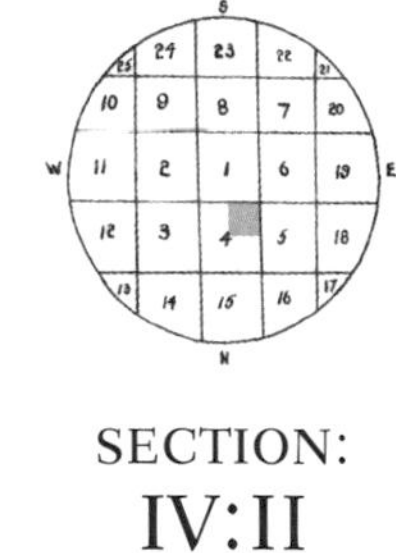

SECTION:
IV:II

Some mountains on the Moon were named after mountains on Earth in the 17th century. Astronomer Johannes Hevelius (1611–87) proposed that the geography of the Moon was similar to the Mediterranean and surrounding lands, and based his lunar names on that region. Some of Hevelius's names are still used: the **Apennines** on this sheet are named after mountains in Italy.

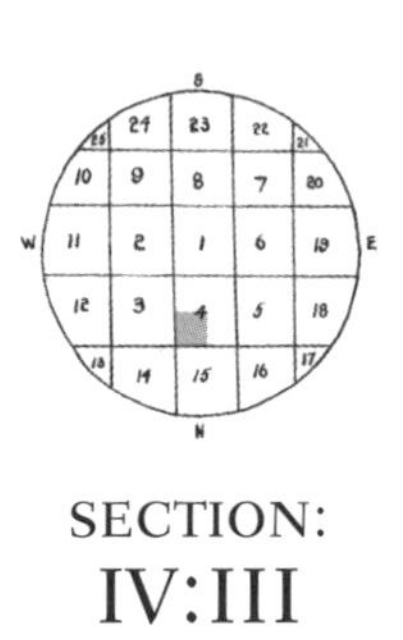

SECTION:
IV:III

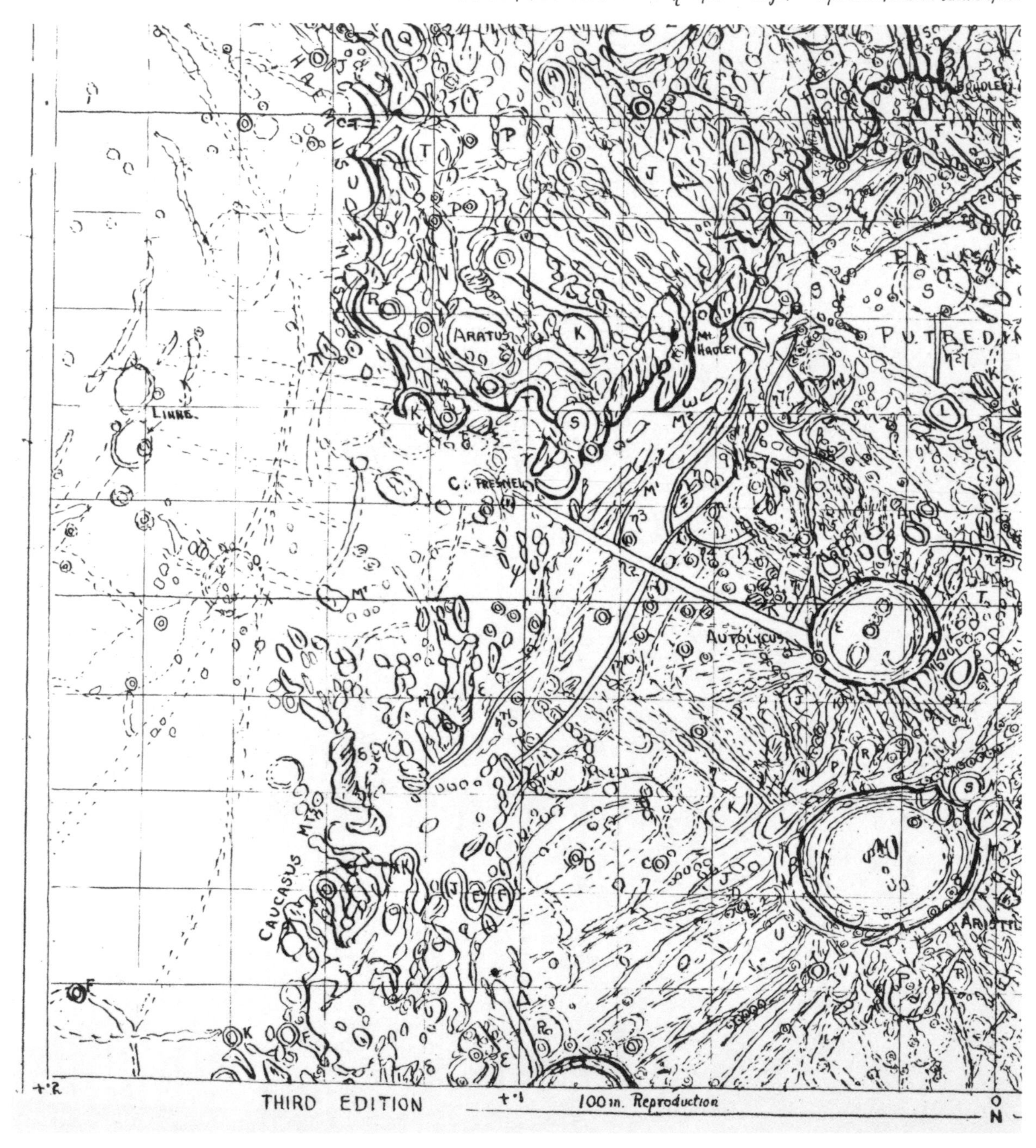

The Apollo 15 landing site (30 July 1971) was just to the side of the crater **Autolycus**.

KEY:- ◎ Craterlet, ⊙ Craterpit, ⊕ Cratercone, ○ Hillock, ⏉ Clefts, ═ Ridges, ······ Light Streaks, ⋕ Dark Variable Spots.

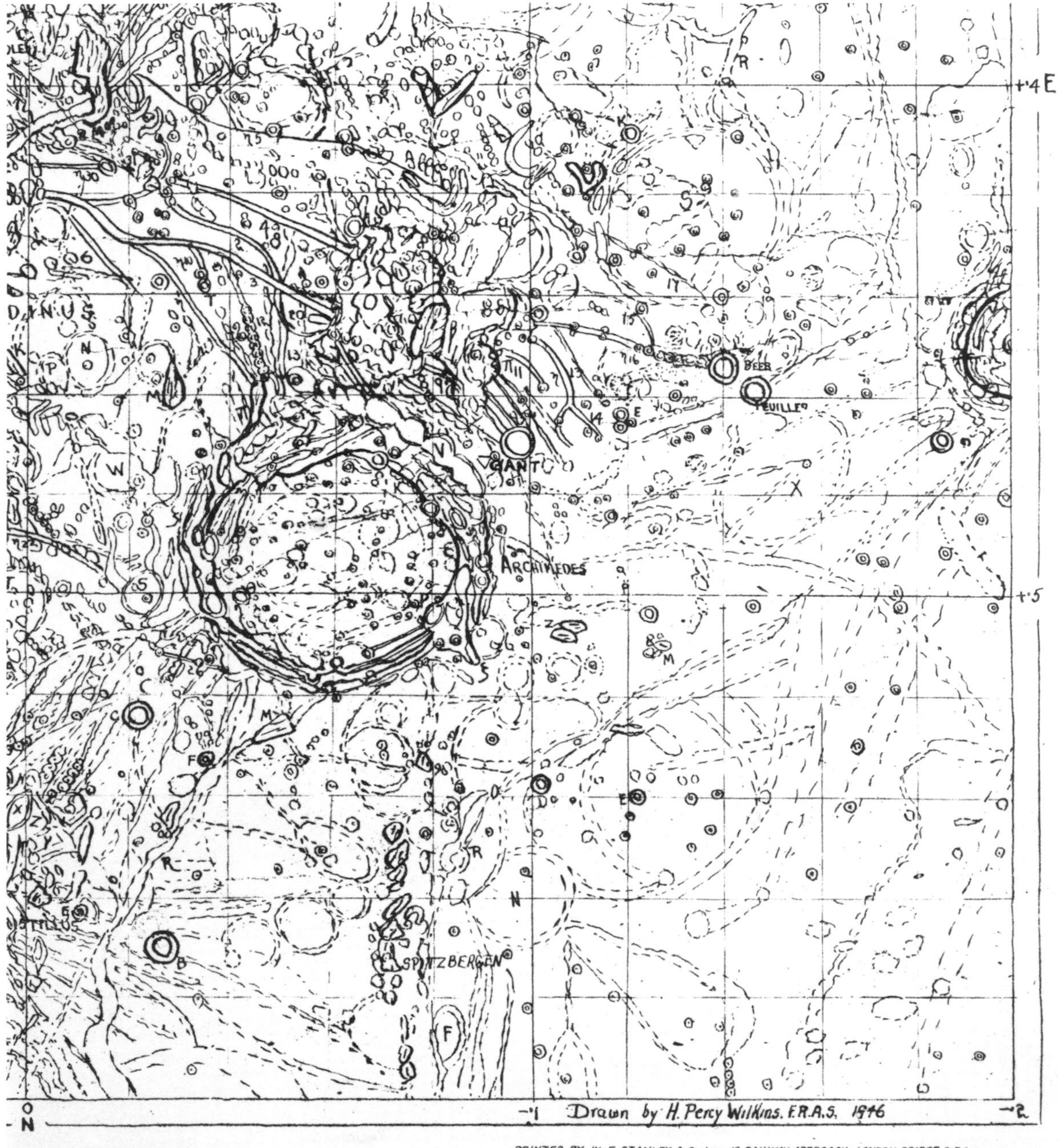

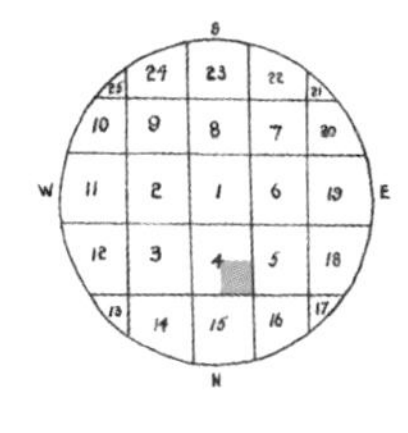

SECTION:
IV:IV

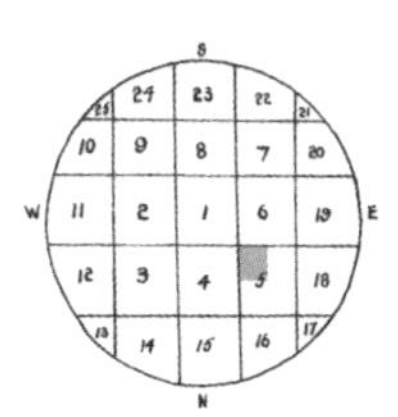

SECTION:
V:I

KEY:- ⊚ Craterlet, ⊙ Craterpit, ⊕ Cratercone, o Hillock, Clefts, ═ Ridges, ······ Light Streaks, Dark Variable Spots.

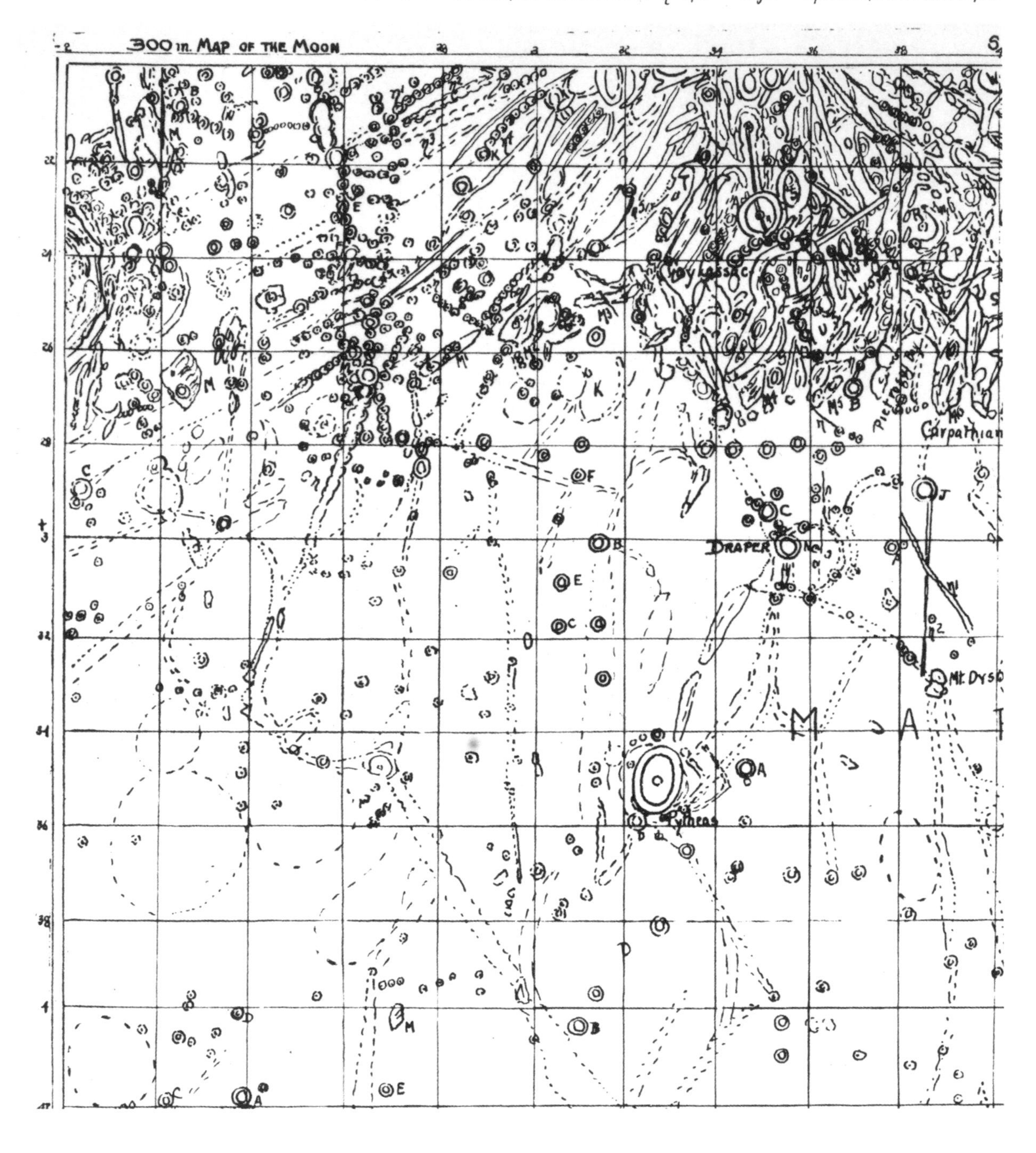

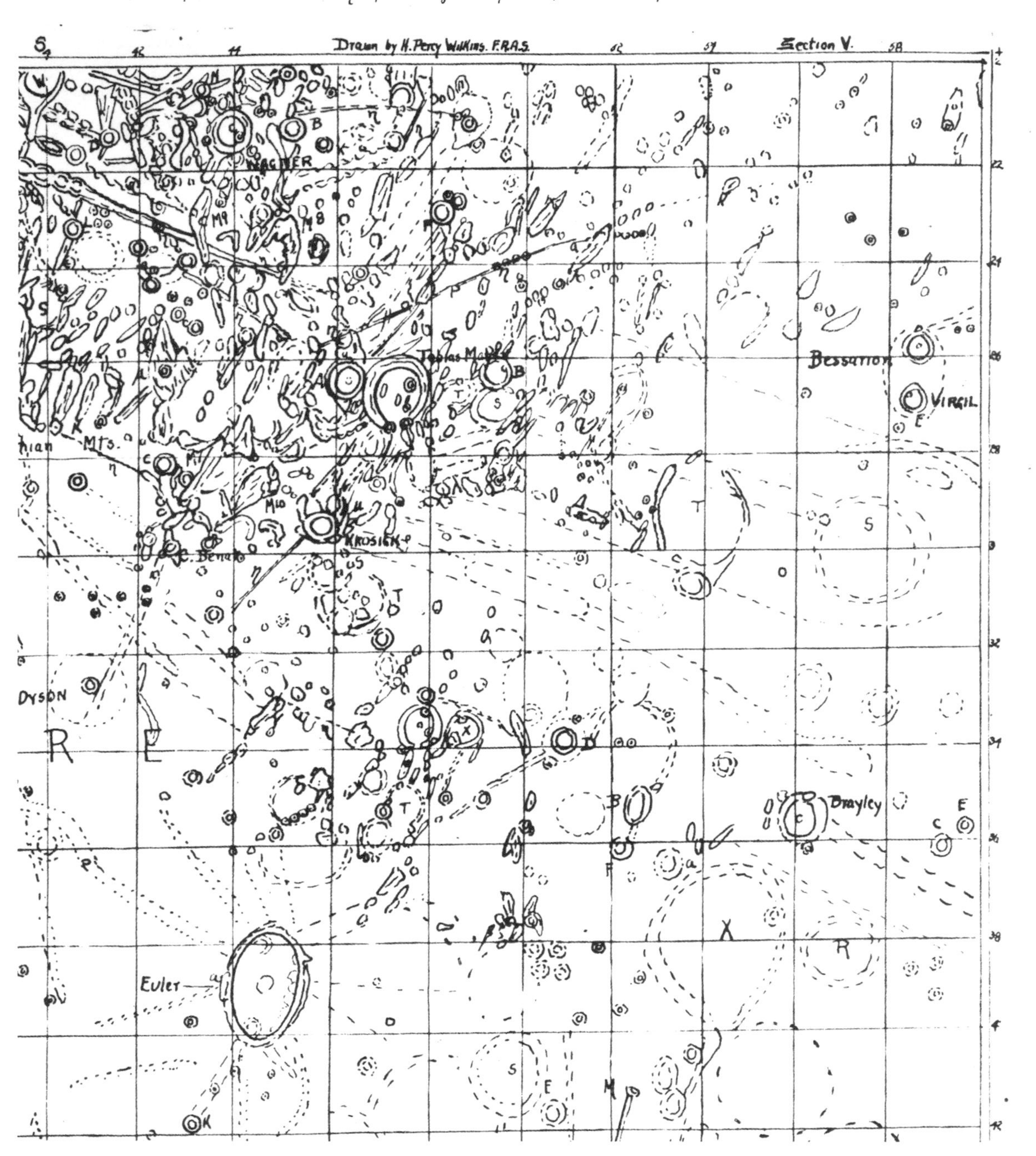
KEY:- ◎ Craterlet, ⊙ Craterpit, ⊕ Cratercone, ○ Hillock, Clefts, ═ Ridges, ······ Light Streaks, Dark Variable Spots.
Drawn by H. Percy Wilkins. F.R.A.S.
Section V.
WAGNER
Tobias Mayer
Bessarion
VIRGIL
KRUSICK
Mts.
DYSON
R
E
Brayley
Euler

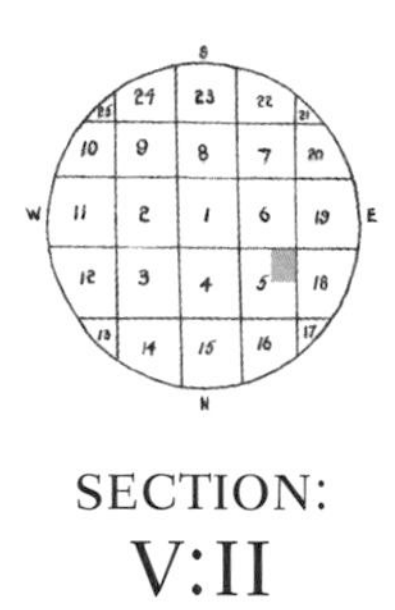

SECTION:
V:II

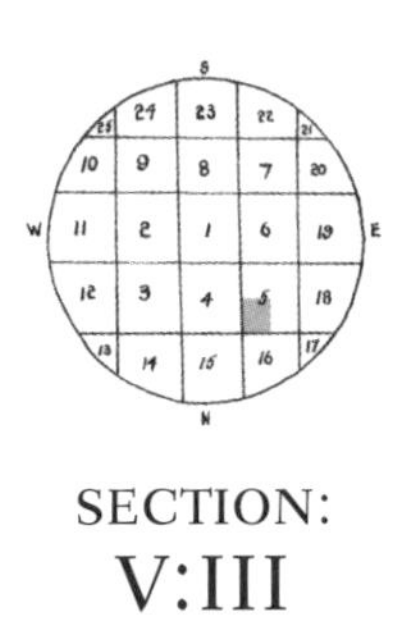

SECTION:
V:III

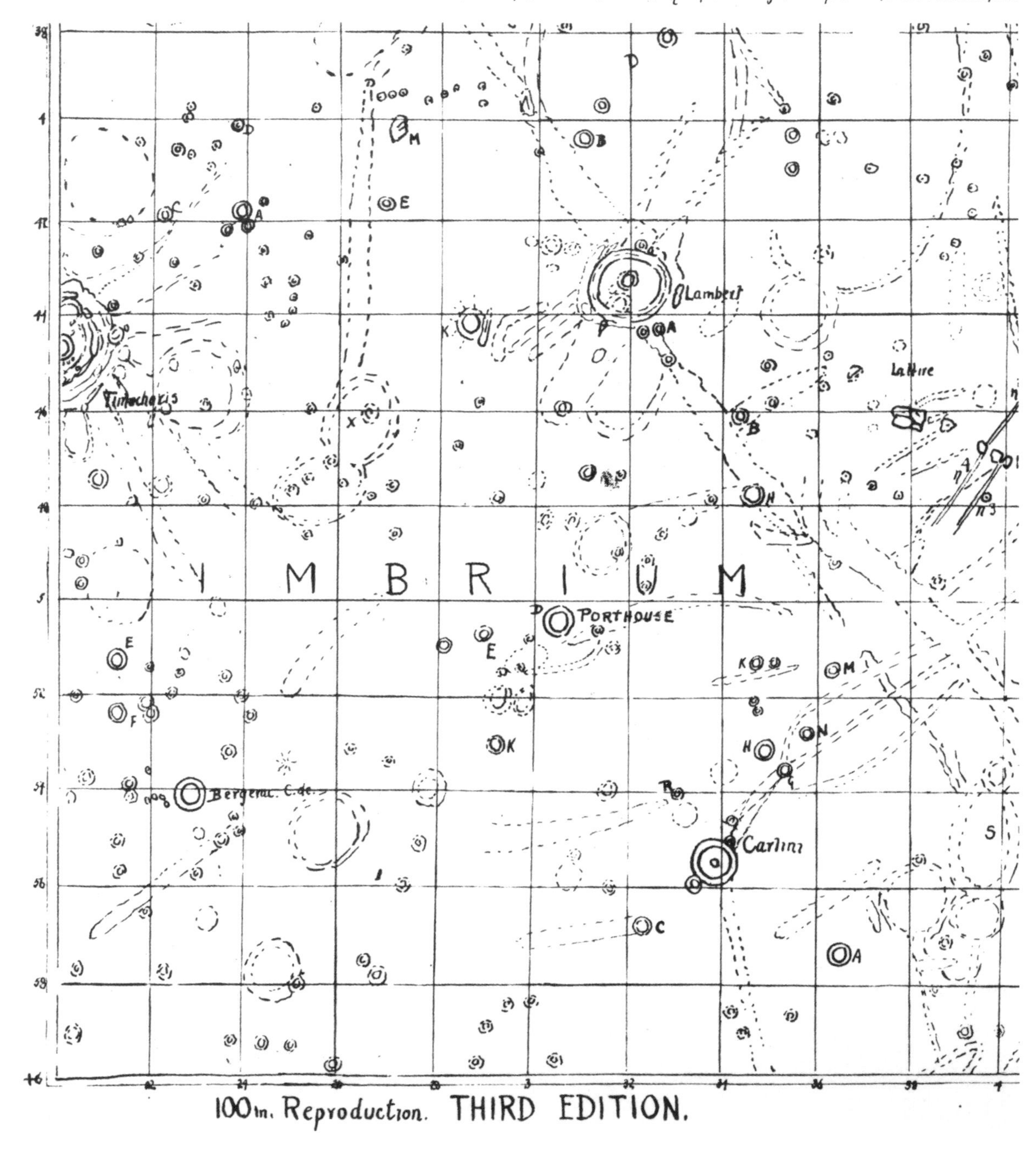
KEY:- Craterlet, Craterpit, Cratercone, Hillock, Clefts, Ridges, Light Streaks, Dark Variable Spots.
IMBRIUM
Lambert
La Hire
Timocharis
PORTHOUSE
Bergerac. C.de.
Carlini
100in. Reproduction. THIRD EDITION.

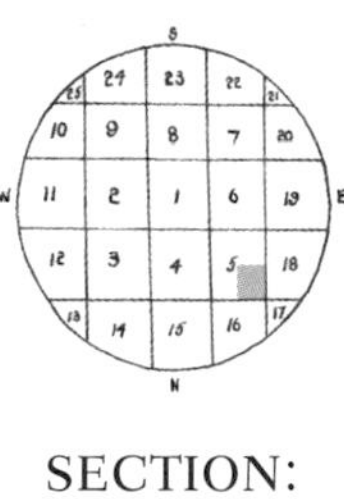

SECTION:
V:IV

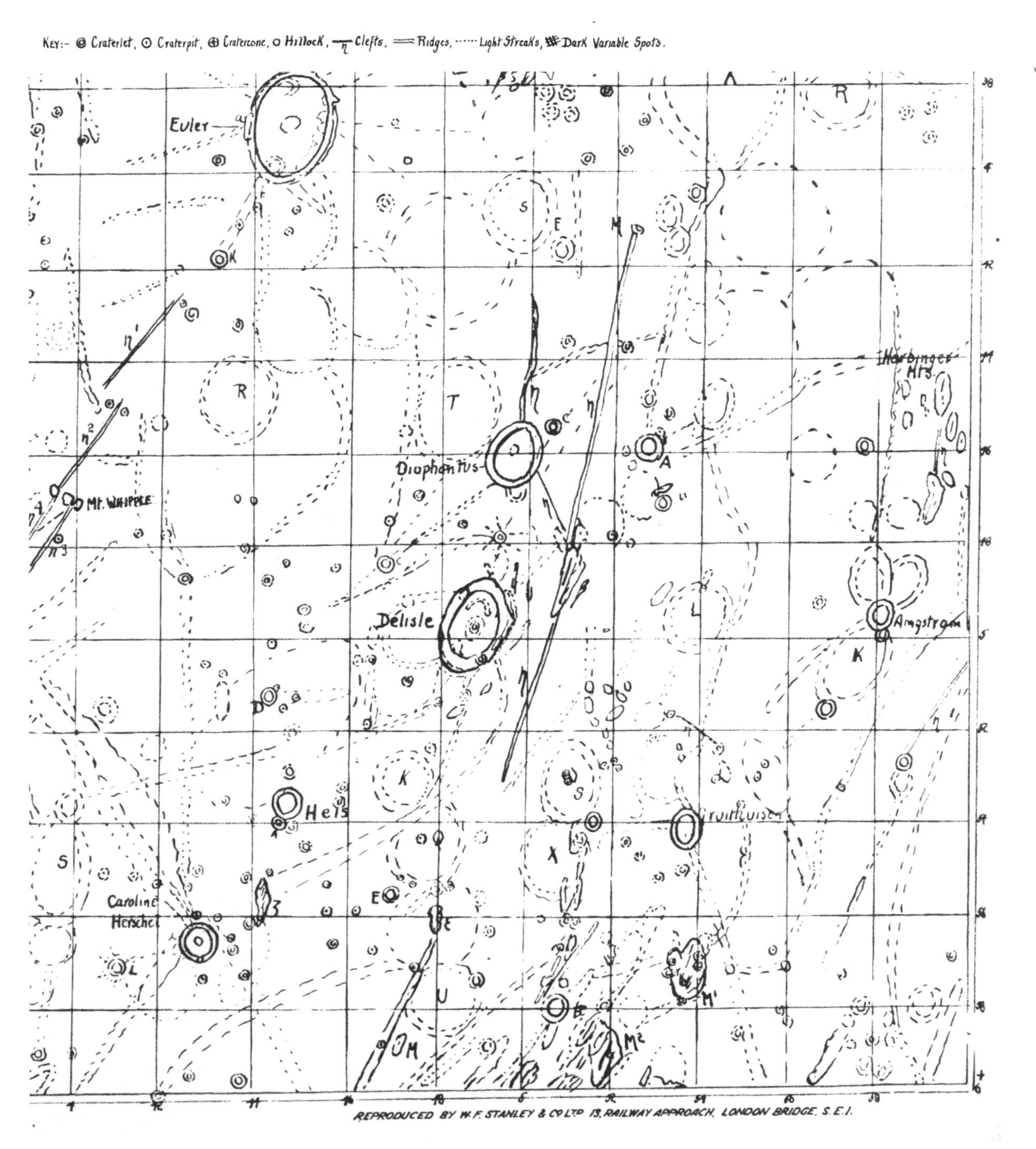

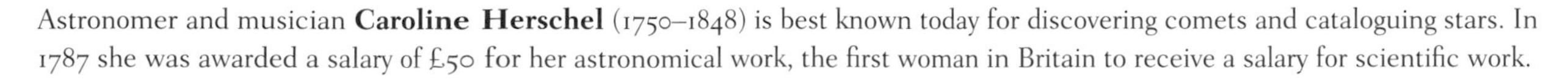

Astronomer and musician **Caroline Herschel** (1750–1848) is best known today for discovering comets and cataloguing stars. In 1787 she was awarded a salary of £50 for her astronomical work, the first woman in Britain to receive a salary for scientific work.

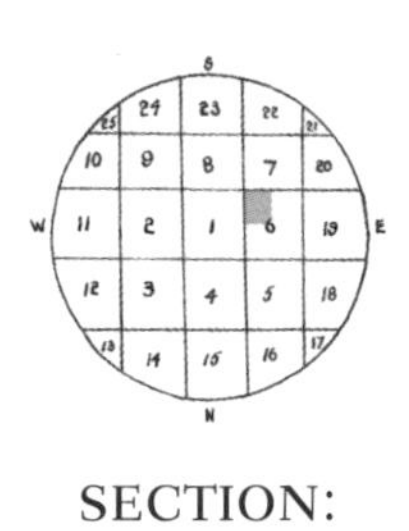

SECTION:
VI:I

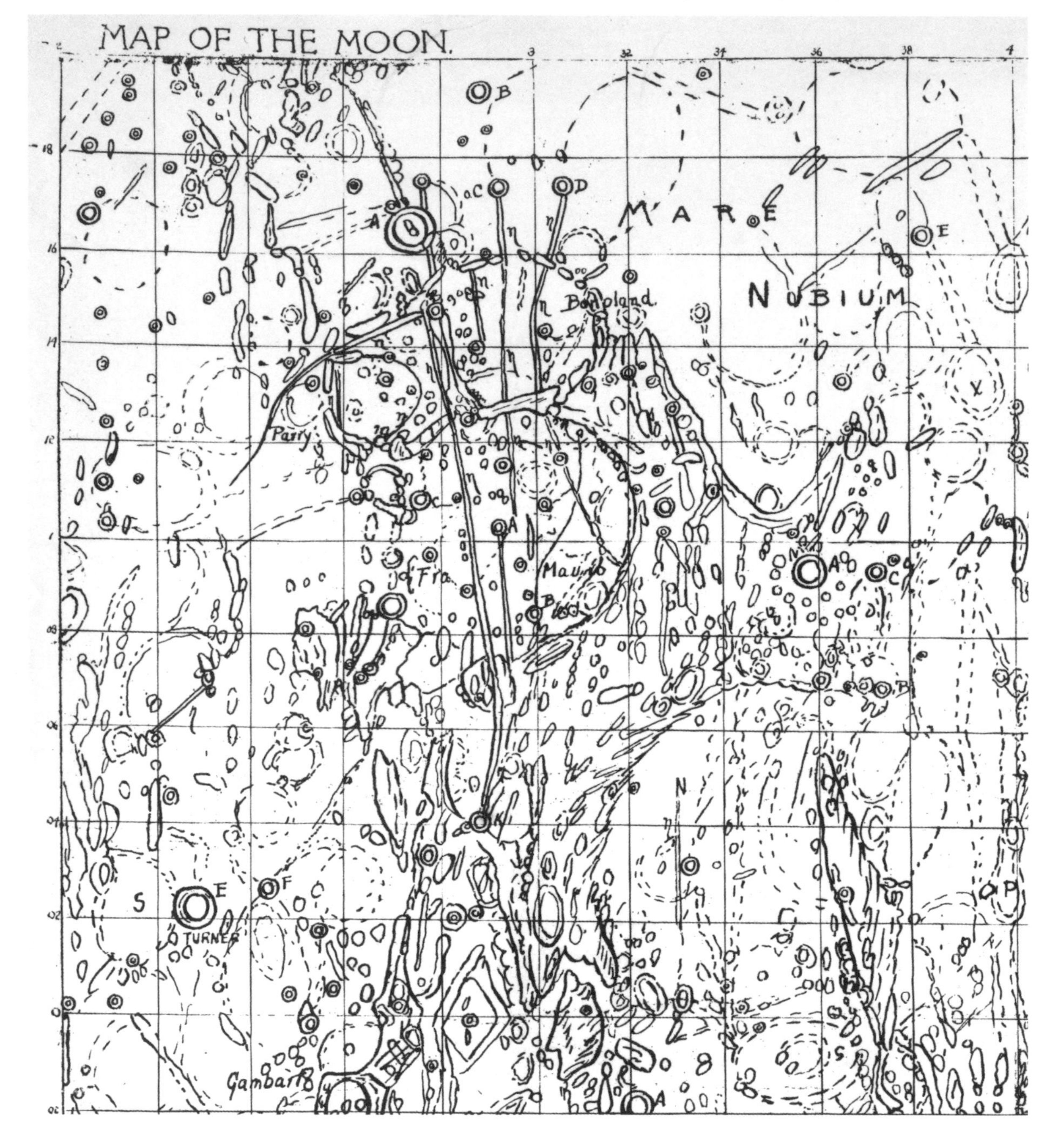
KEY:- Craterlet, Craterpit, Cratercone, Hillock, Clefts, Ridges, Light Streaks, Dark Variable Spots.
MAP OF THE MOON.
MARE
NUBIUM
Bonpland
Parry
Fra
Mauro
TURNER
Gambart

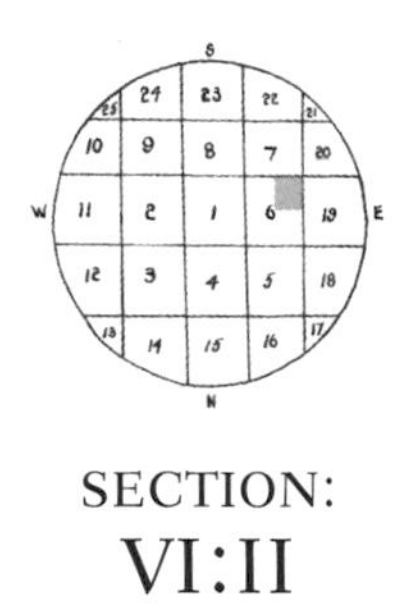

SECTION:
VI:II

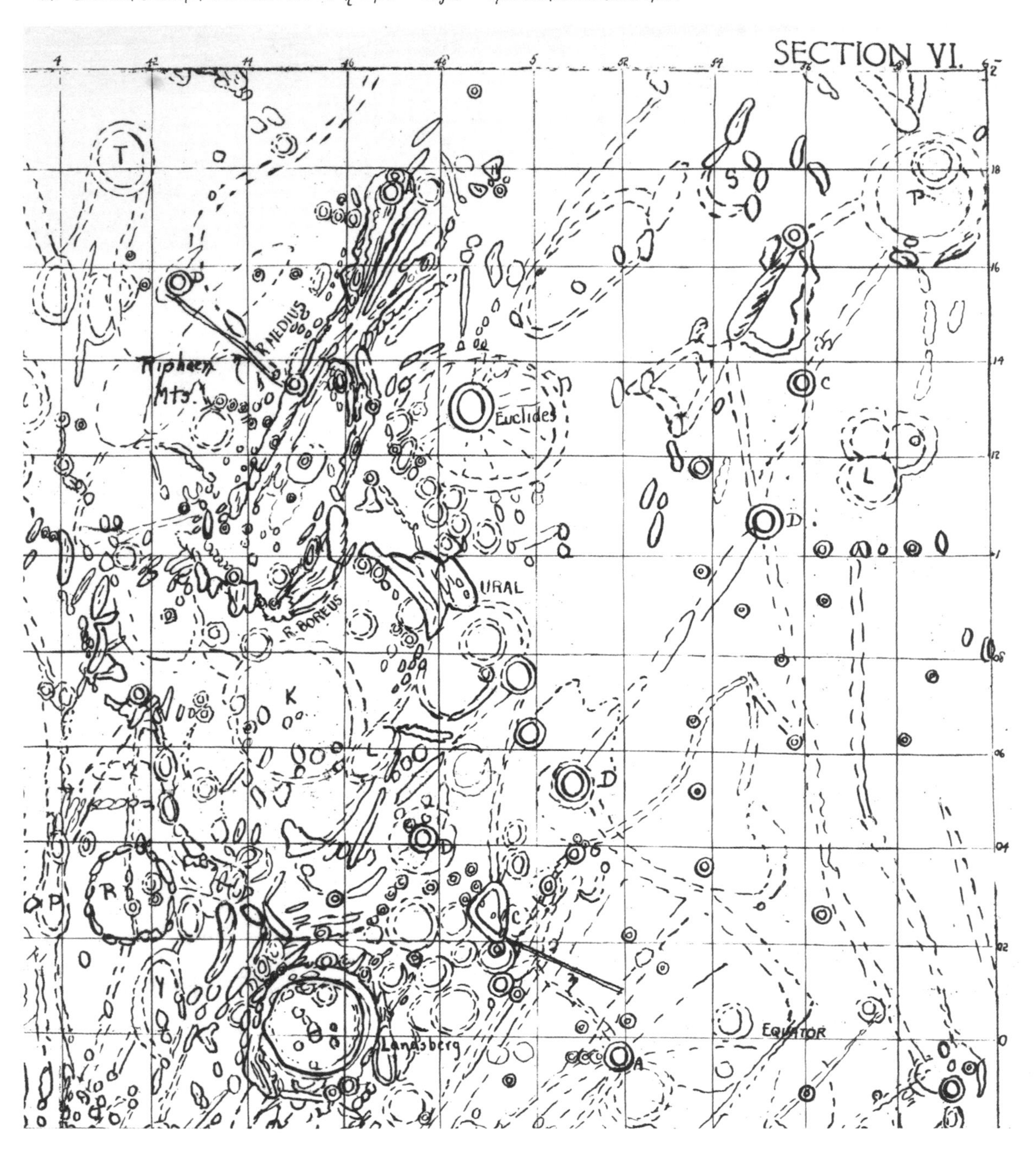

KEY:- Craterlet, Craterpit, Cratercone, Hillock, Clefts, Ridges, Light Streaks, Dark Variable Spots.
SECTION VI.
Euclides
Landsberg
URAL
R. BOREUS
EQUATOR

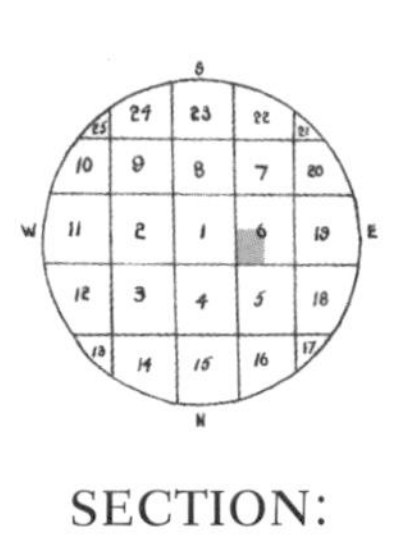

SECTION: VI:III

KEY:- Craterlet, Craterpit, Cratercone, Hillock, Clefts, Ridges, Light Streaks, Dark Variable Spots.

Gambart

Reinhold

Stadius

300 in. MAP OF THE MOON Section VI

THIRD EDITION.

The large crater at the bottom of this sheet is known as **Copernicus**. It is quite a new crater (only 800 million years old), so is very distinct because its rim hasn't been damaged by later asteroid impacts. It is big enough to be seen from Earth using binoculars.

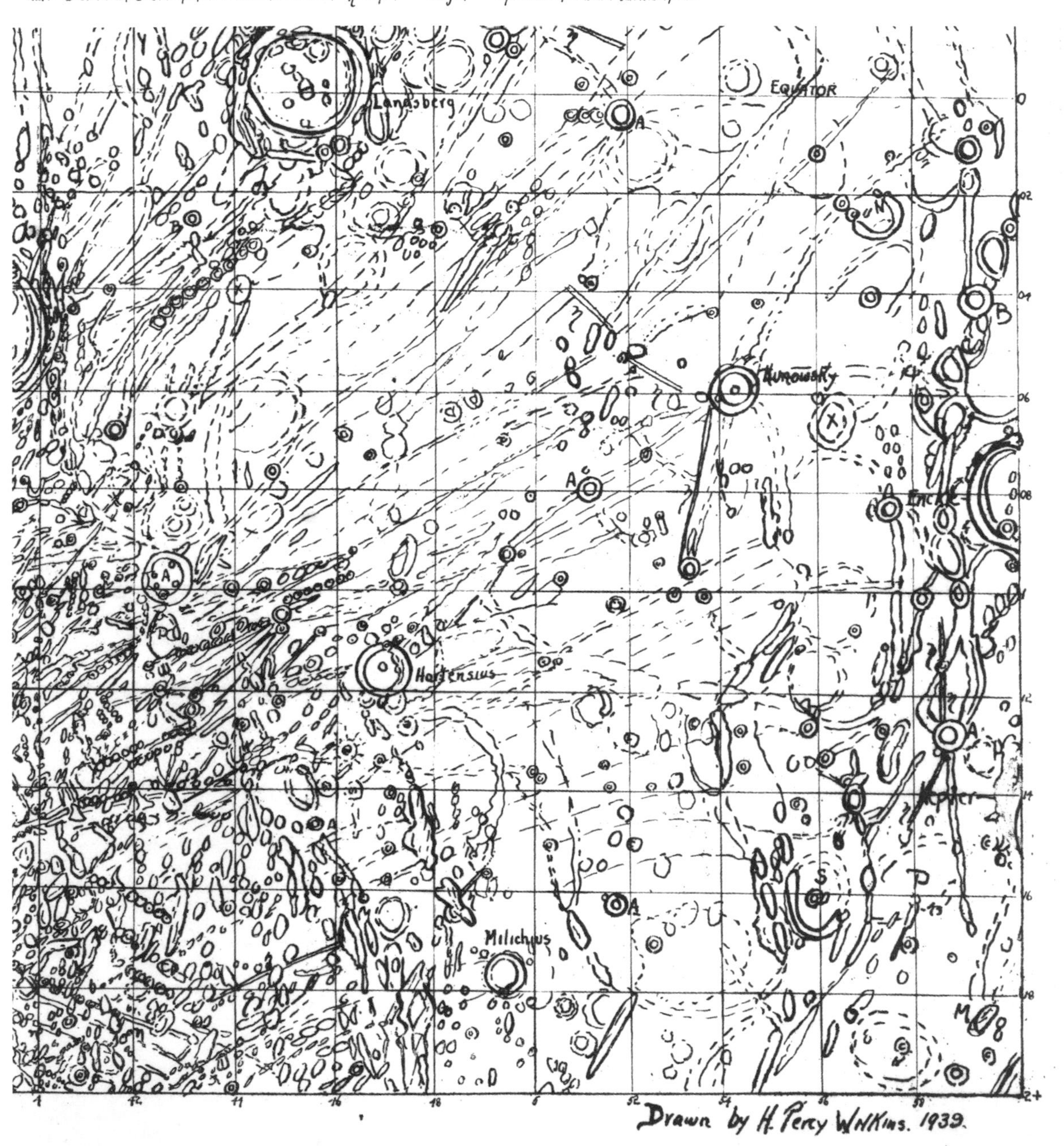

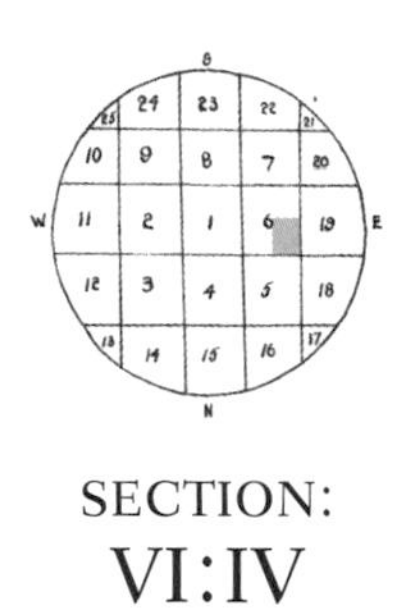

SECTION:
VI:IV

The Apollo 12 landing site (19 November 1969) was just between the craters **Lansberg** (above), **Reinhold** and **Gambart** (page 28).

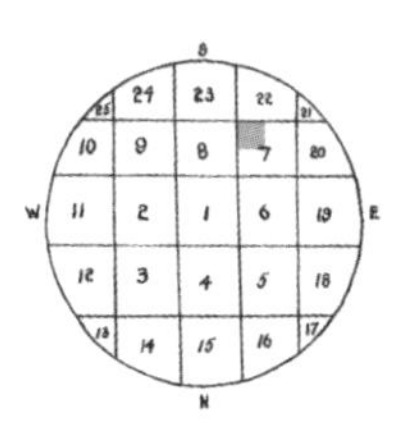

SECTION:
VII:I

KEY:- Craterlet, Craterpit, Cratercone, Hillock, Clefts, Ridges, Light Streaks, Dark Variable Spots.

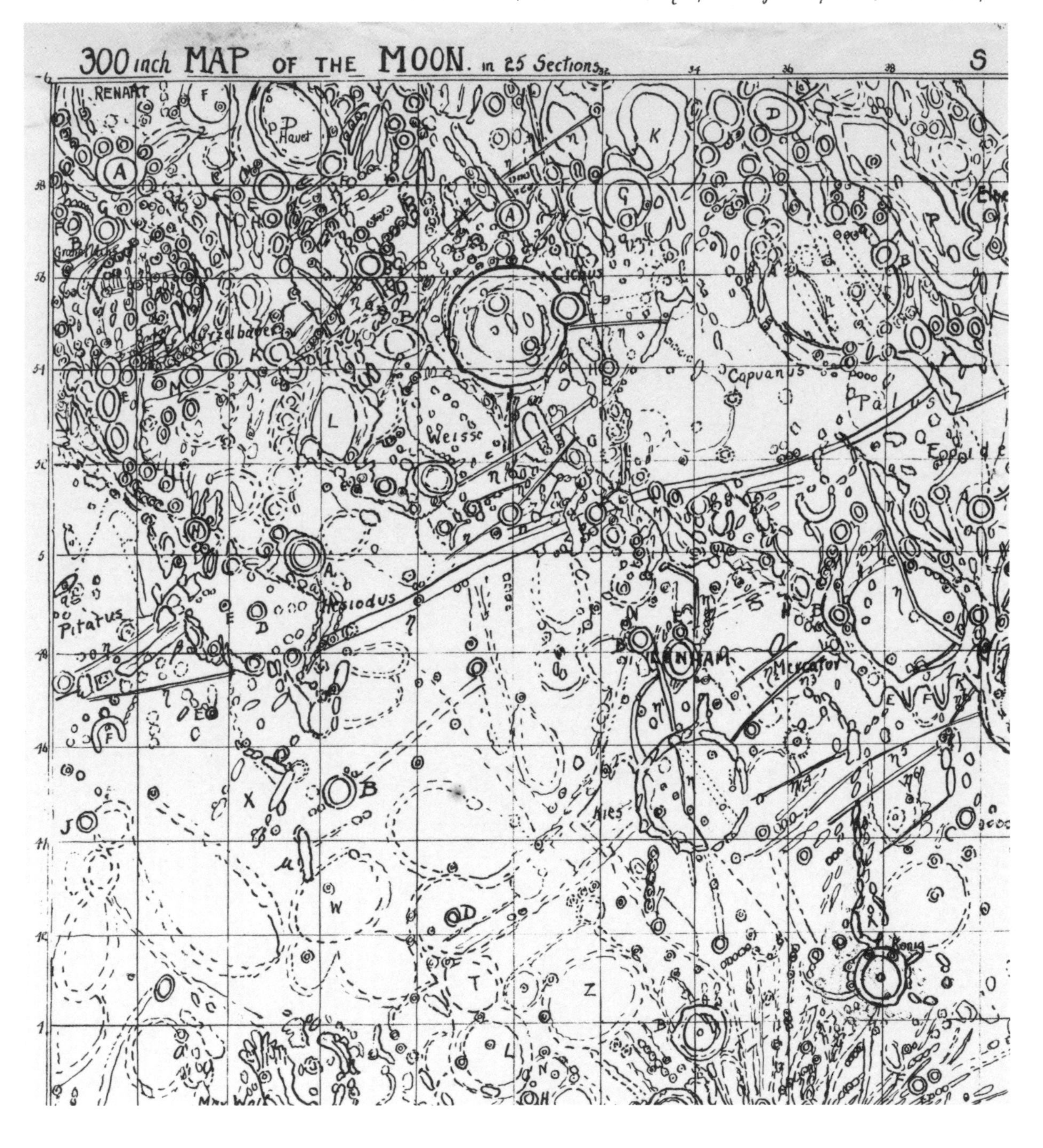

Key:- Craterlet, Craterpit, Cratercone, Hillock, Clefts, Ridges, Light Streaks, Dark Variable Spots.

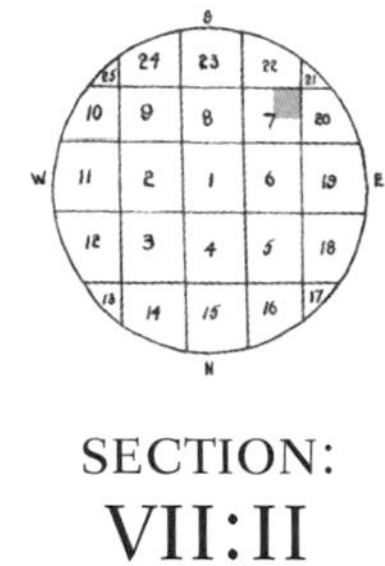

SECTION:
VII:II

While making his Map of the Moon, Wilkins gave new names to various features, unauthorised by the International Astronomical Union. Several of these were the names of people he had worked with, including **Patrick Moore**, a fellow amateur astronomer with whom Wilkins co-authored a book about the Moon. Moore later became famous presenting the BBC's *The Sky at Night*.

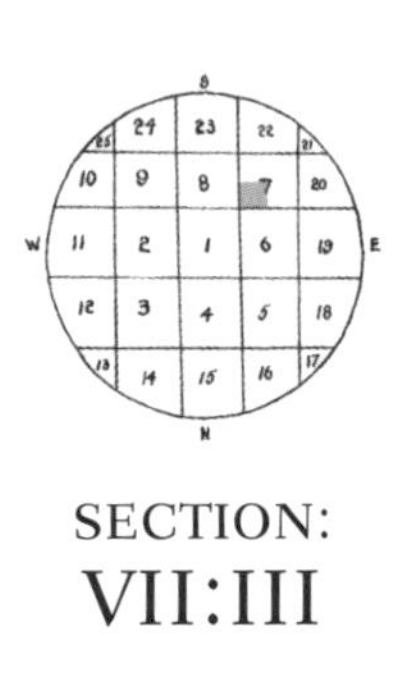

SECTION:
VII:III

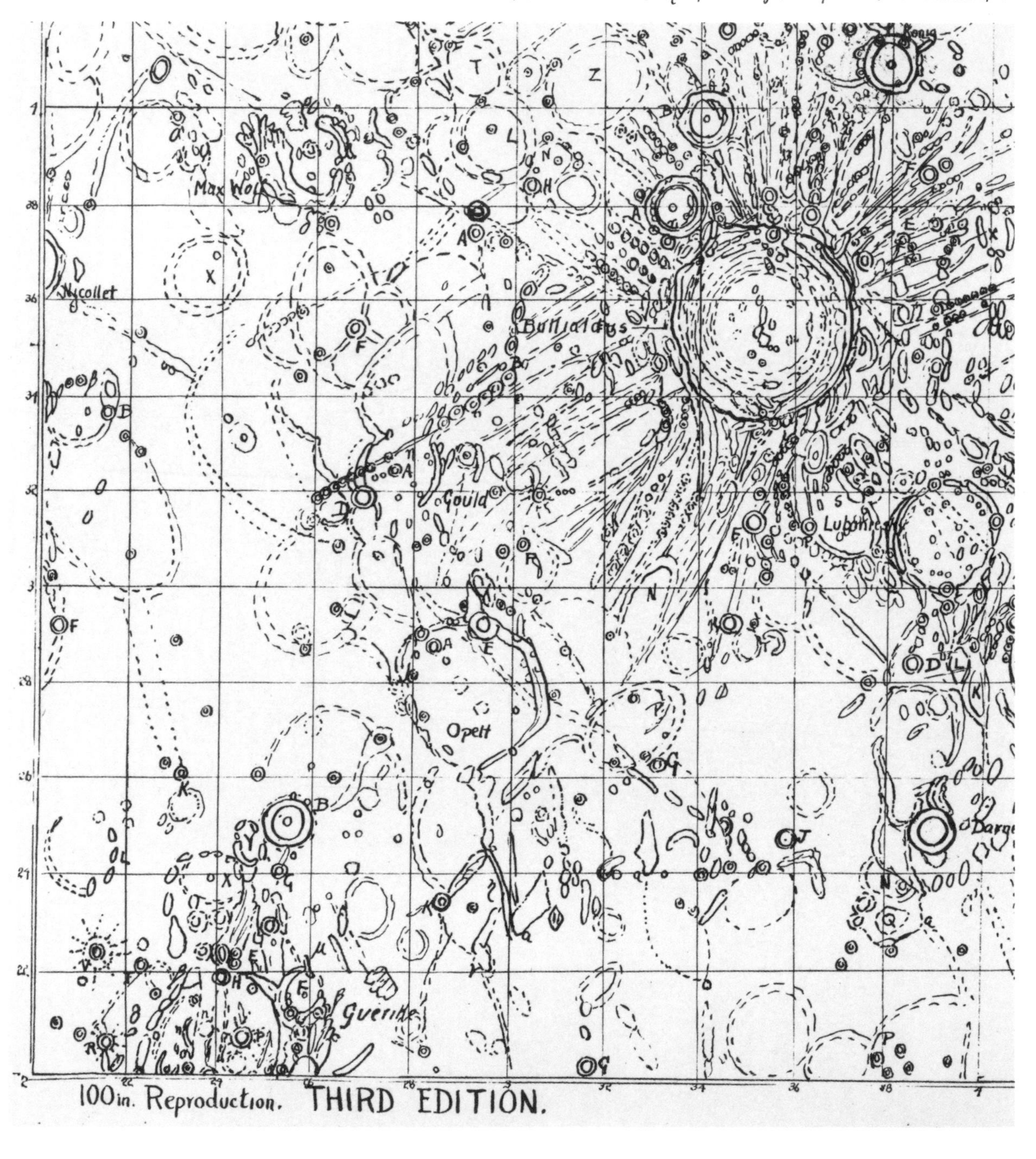
KEY:- Craterlet, Craterpit, Cratercone, Hillock, Clefts, Ridges, Light Streaks, Dark Variable Spots.
Max Wolf
Nycollet
Bullialdus
Gould
Lubiniezky
Opelt
Guerike
100in. Reproduction. THIRD EDITION.

KEY:- ◎ Craterlet, ⊙ Craterpit, ⊕ Cratercone, o Hillock, Clefts, ══ Ridges, ······ Light Streaks, Dark Variable Spots.

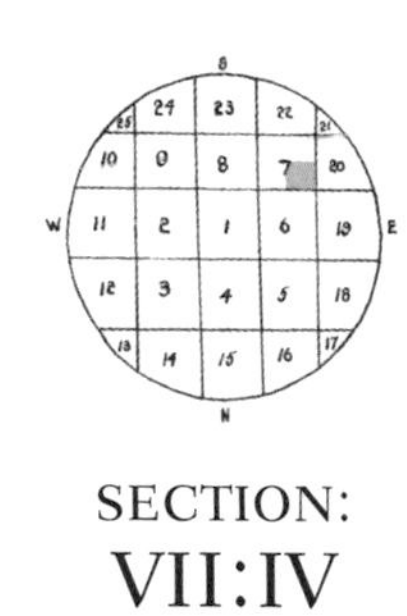

SECTION:
VII:IV

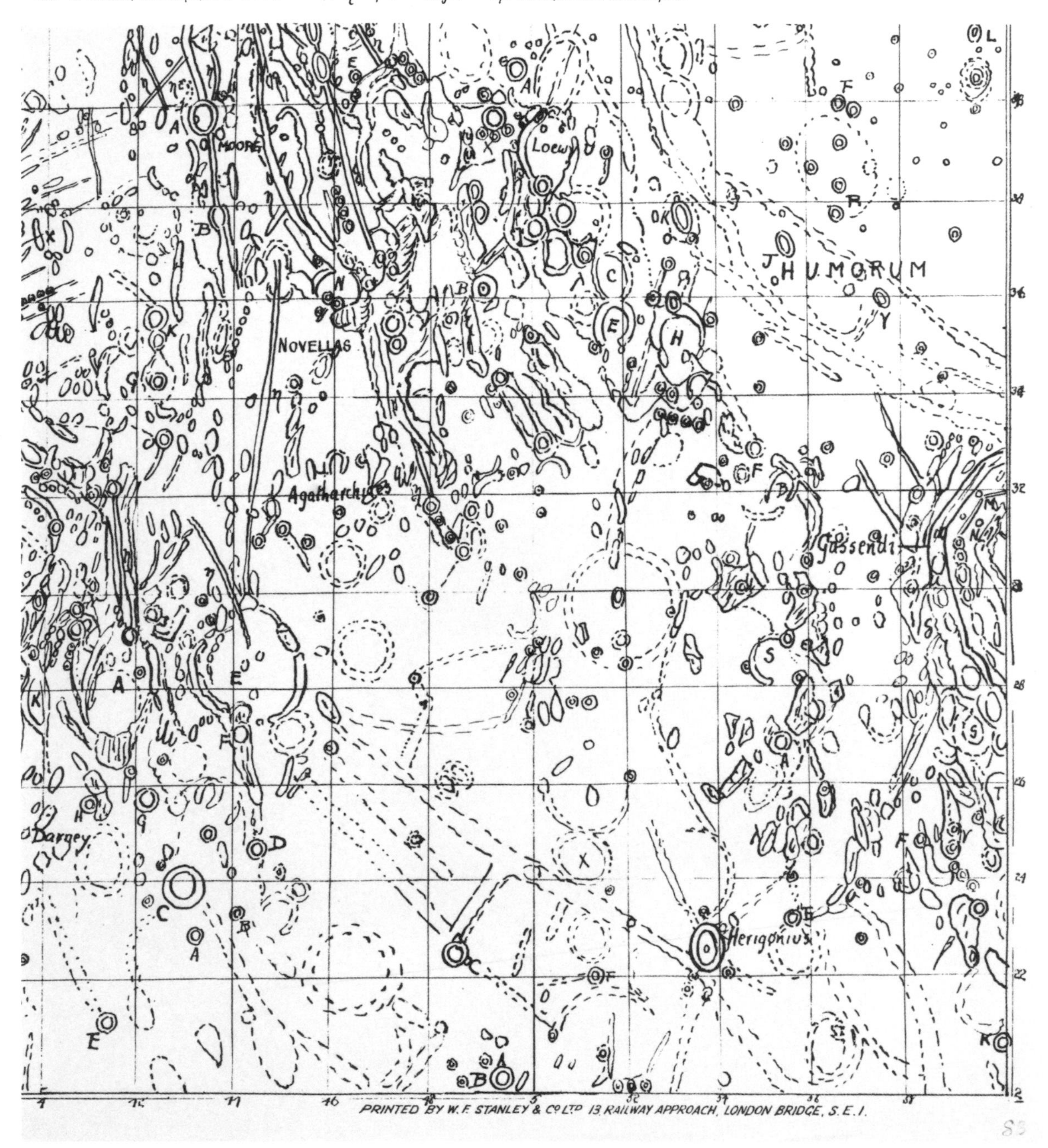

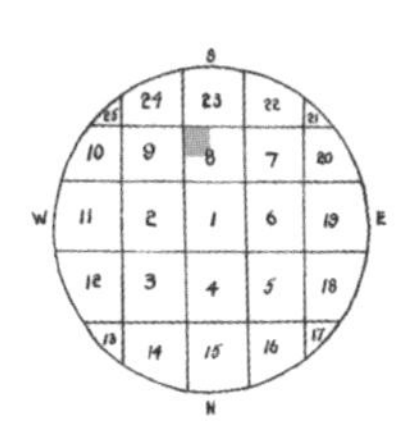

SECTION:
VIII:I

KEY:- ◎ Craterlet, ⊙ Craterpit, ⊕ Cratercone, o Hillock, Clefts, Ridges, ····· Light Streaks, Dark Variable Spots.

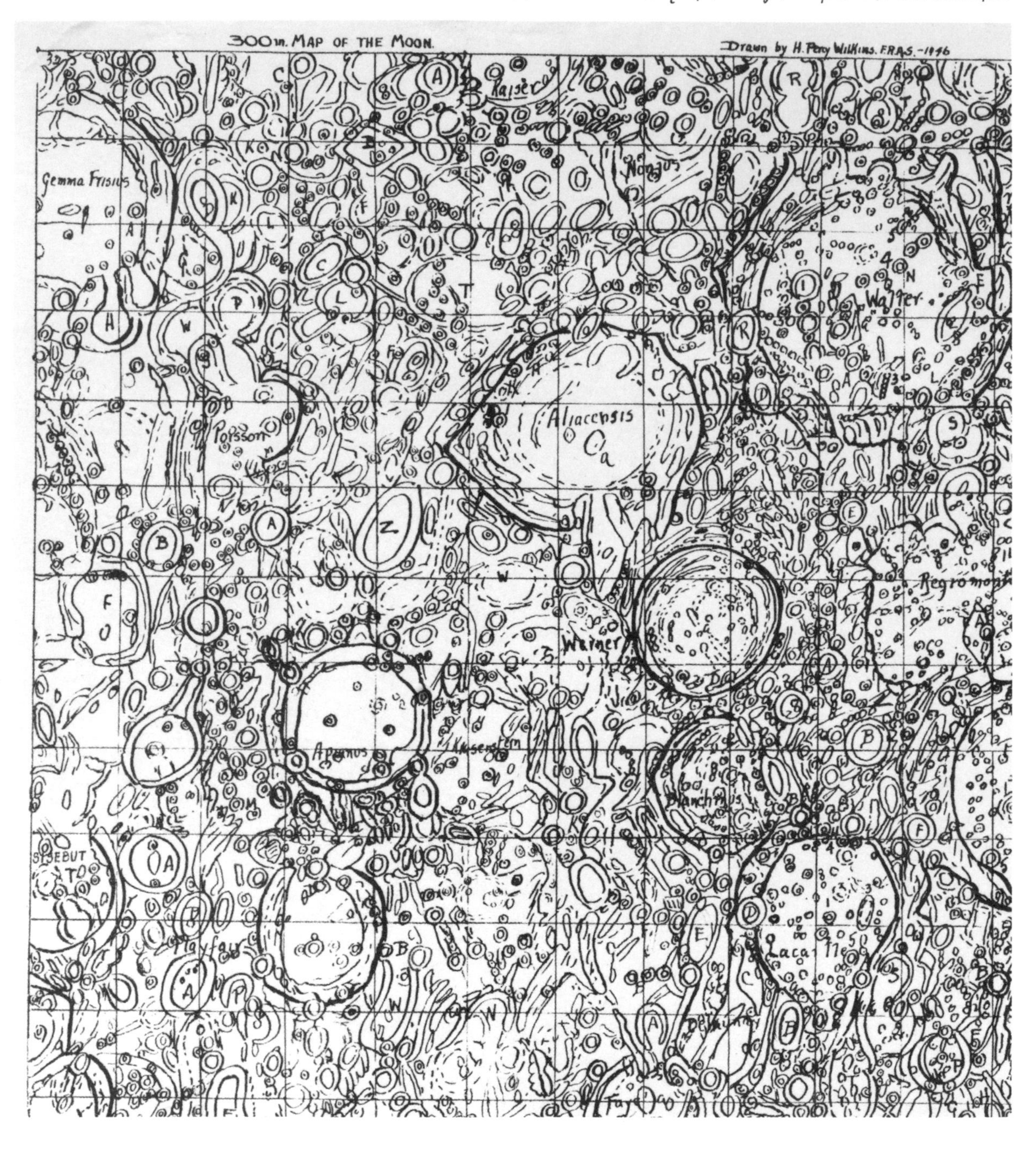

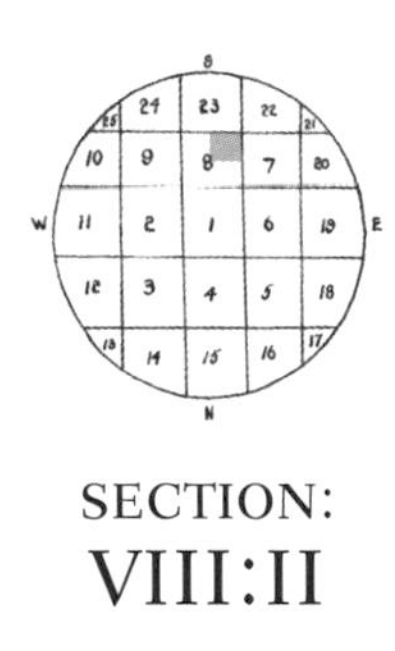

SECTION: VIII:II

KEY:- ◎ Craterlet, ⊙ Craterpit, ⊕ Cratercone, ○ Hillock, Clefts, ═ Ridges, ... Light Streaks, Dark Variable Spots.

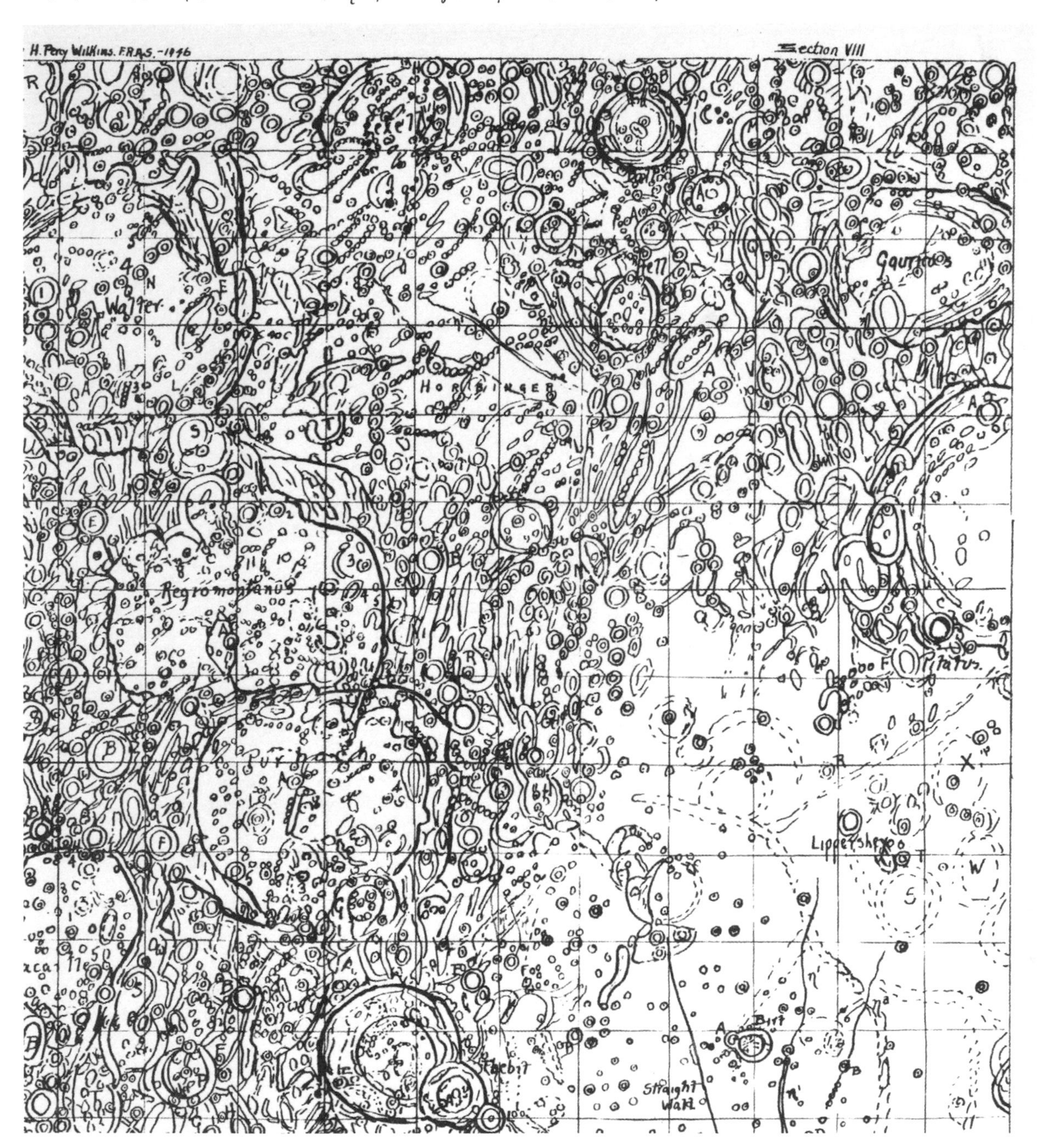

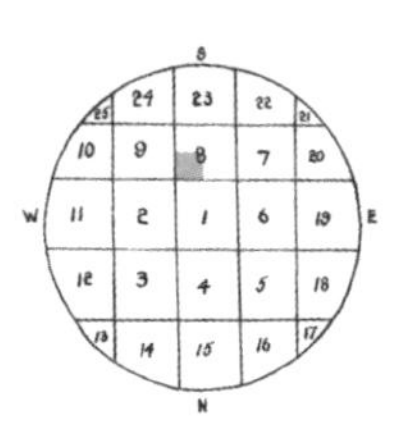

SECTION:
VIII:III

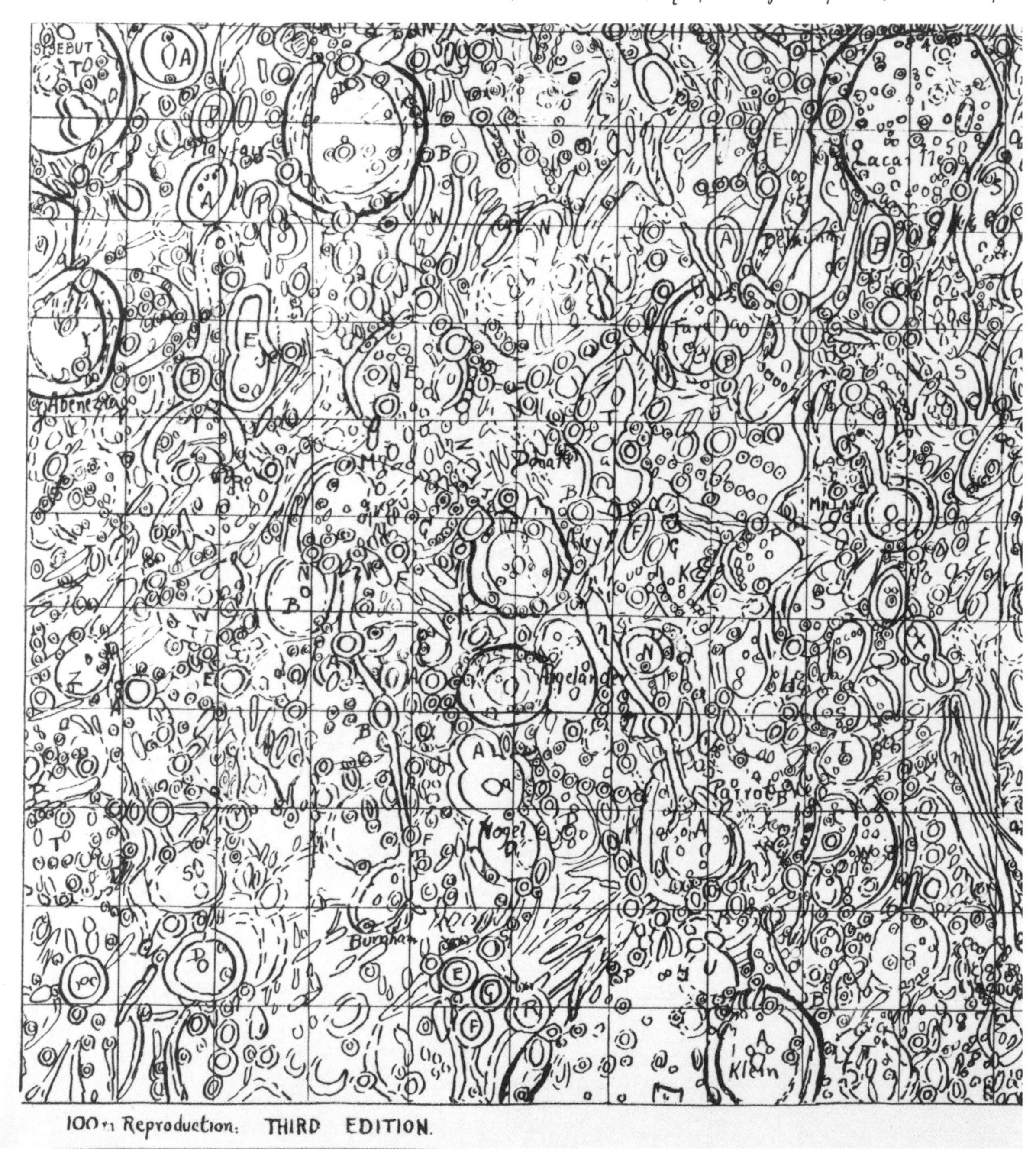

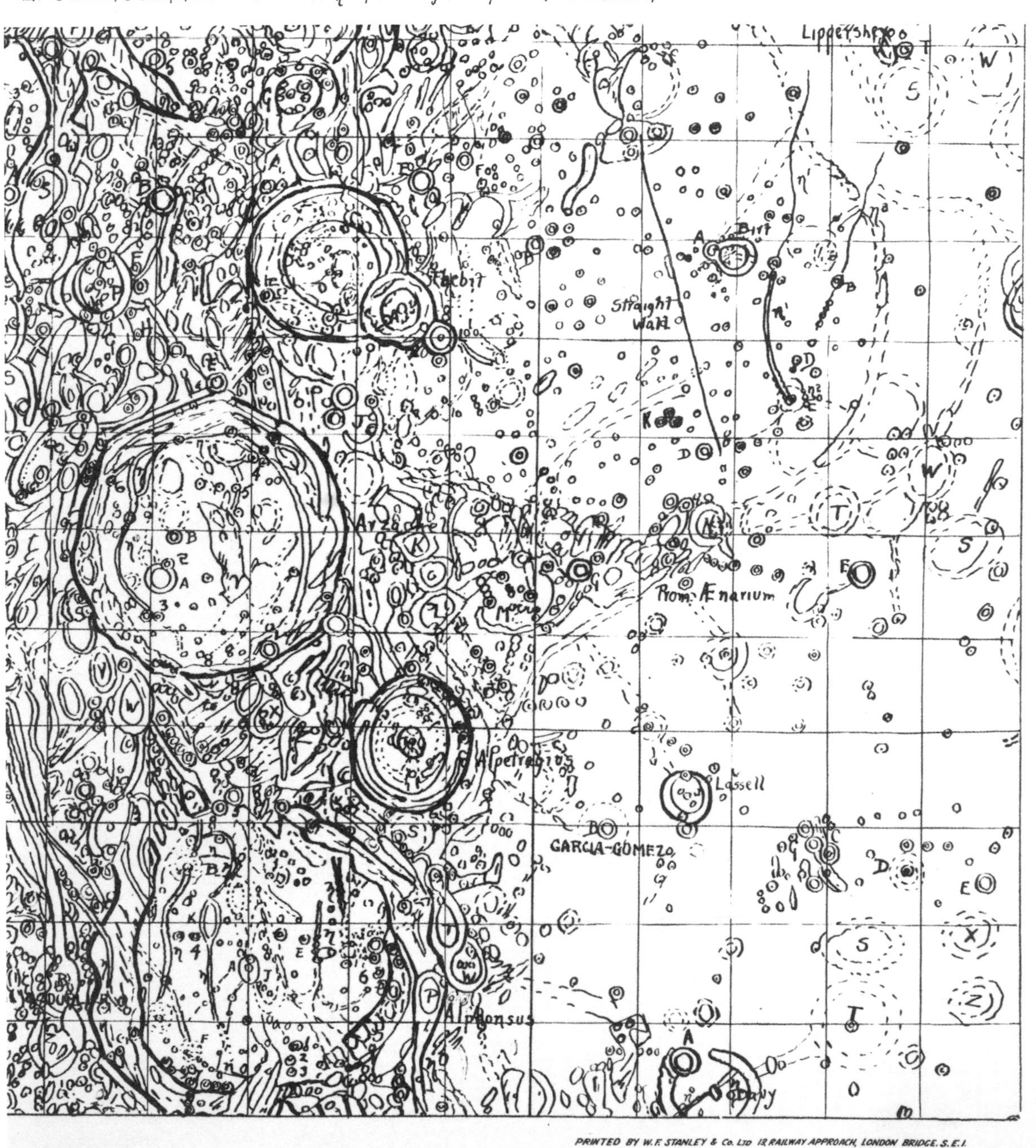

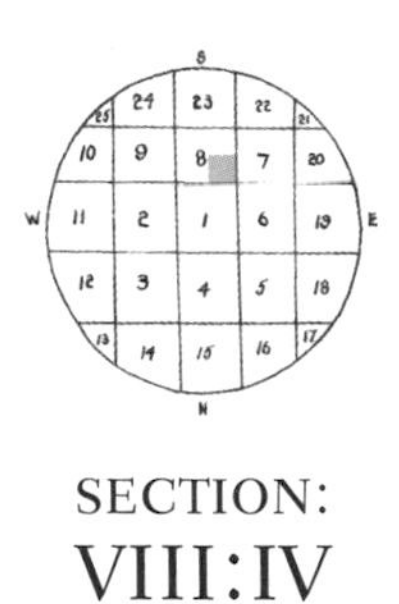

SECTION:
VIII:IV

Straight Wall is a linear fault (a line caused by a fracture in the Moon's crust) on the surface of the Moon. It is about 68 miles long and between 250 and 300 metres high. Its distinctive appearance led Christiaan Huygens, a 17th century scholar, to describe it as a sword, with Straight Wall as the blade and the craters to the south forming the handle.

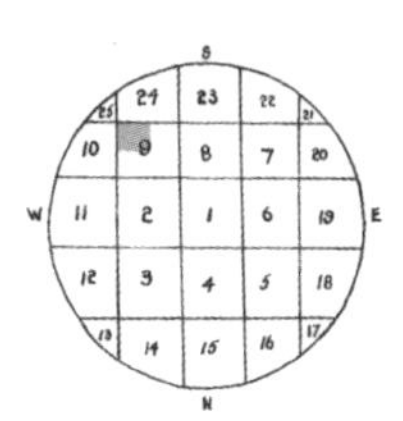
8
24
23
22
10
9
8
7
20
W
11
2
1
6
19
E
12
3
4
5
18
14
15
16
N

SECTION:
IX:I

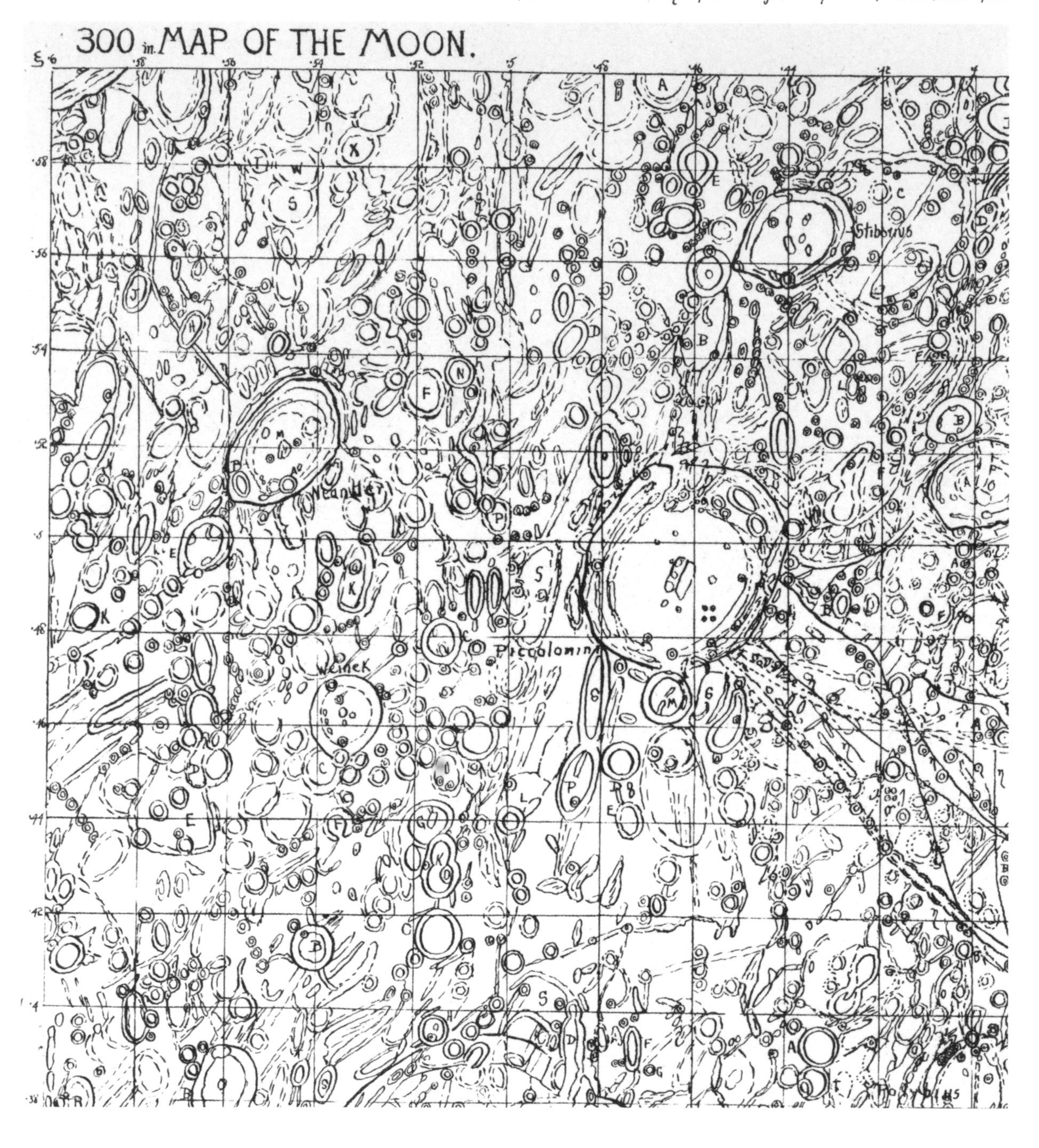
KEY:- ⊚ Craterlet, ⊙ Craterpit, ⊕ Cratercone, ○ Hillock, Clefts, Ridges, Light Streaks, Dark Variable Spots.
300 in. MAP OF THE MOON.
Stiborius
Neander
Piccolomini
Weinek

KEY:- ⊚ Craterlet, ⊙ Craterpit, ⊕ Cratercone, o Hillock, Clefts, ═══ Ridges, ······ Light Streaks, Dark Variable Spots.

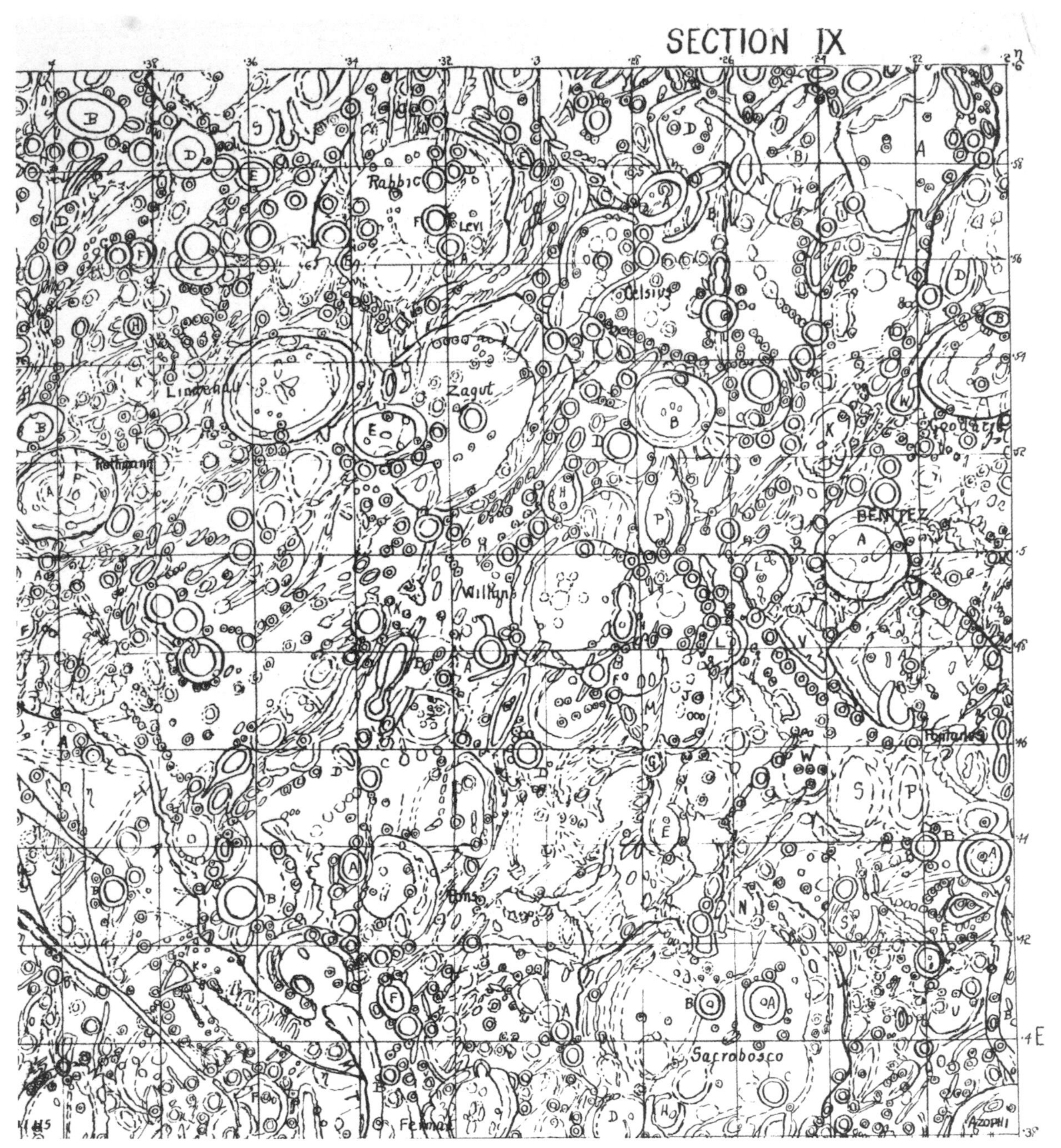

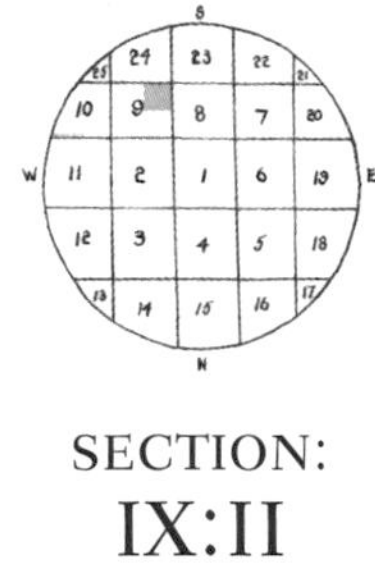

SECTION:
IX:II

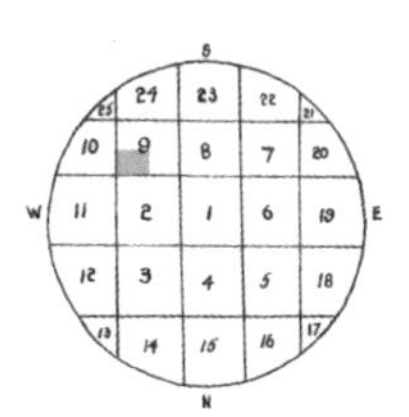

SECTION:
IX:III

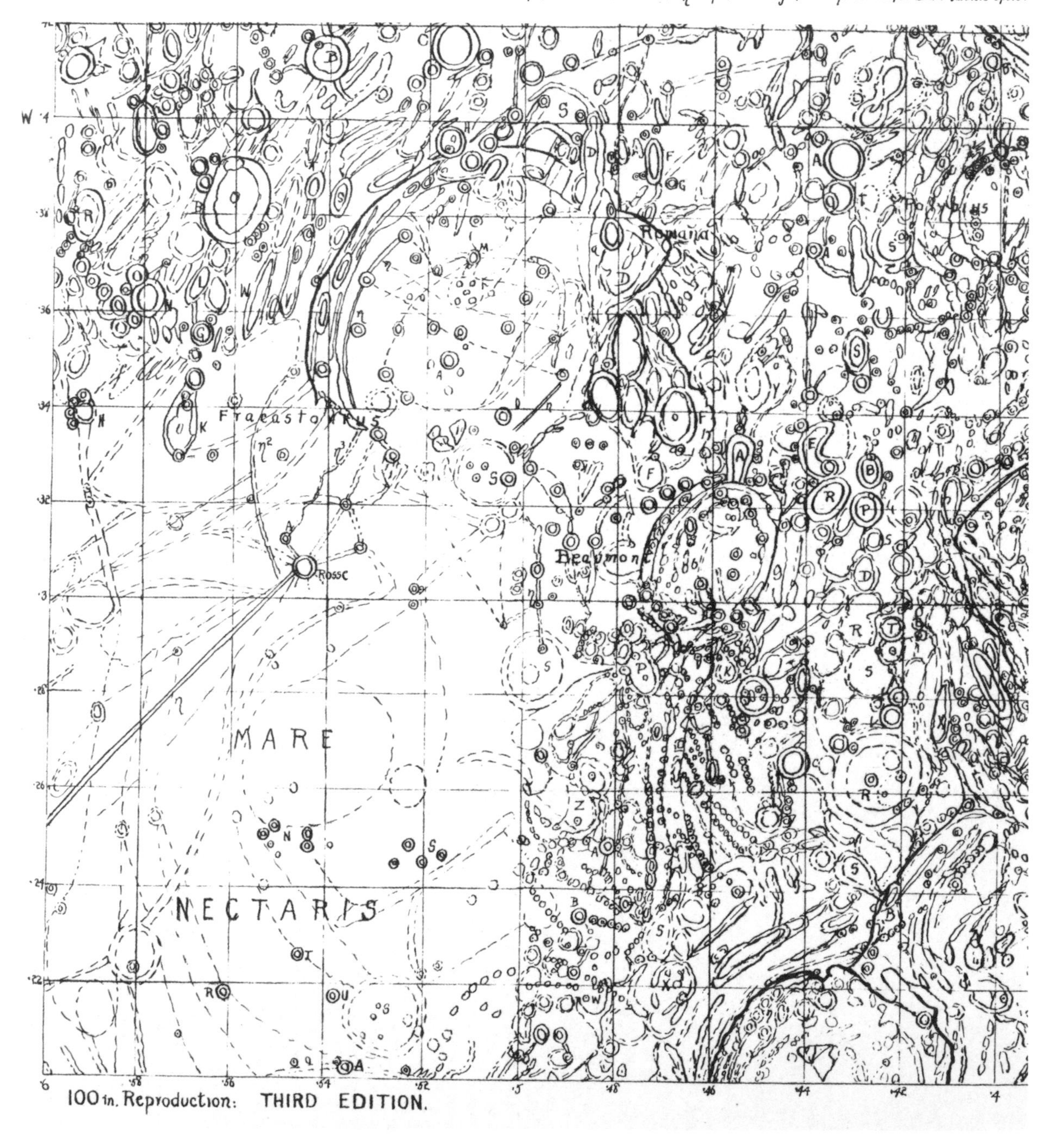
KEY:- ⊚ Craterlet, ⊙ Craterpit, ⊕ Cratercone, ○ Hillock, Clefts, ═ Ridges, ······ Light Streaks, Dark Variable Spots.
Fracastorius
Romana
Beaumont
Rosse
MARE
NECTARIS
100 in. Reproduction: THIRD EDITION.

KEY:- ◎ Craterlet, ⊙ Craterpit, ⊕ Cratercone, ○ Hillock, Clefts, ═ Ridges, ······ Light Streaks, Dark Variable Spots.

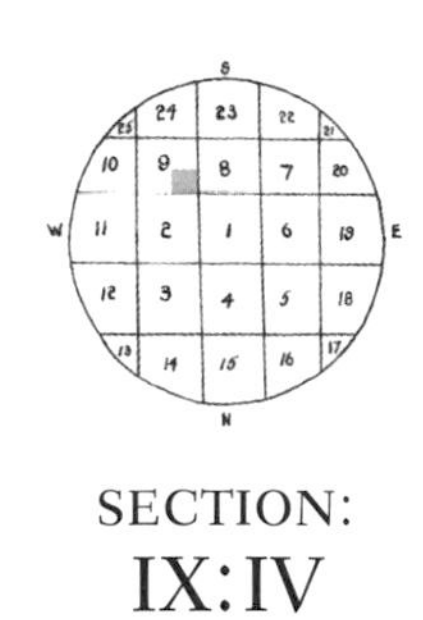

SECTION:
IX:IV

Drawn by H. PERCY WILKINS. F.R.A.S., 1951.

PRINTED BY W.F. STANLEY & Co LTD 13 RAILWAY APPROACH, LONDON BRIDGE S.E.1.

The crater **Azophi** was named in 1651, after the Persian astronomer Abd al-Rahman al-Sufi. His *Book of the Fixed Stars* was one of the most important astronomical texts written in the medieval period.

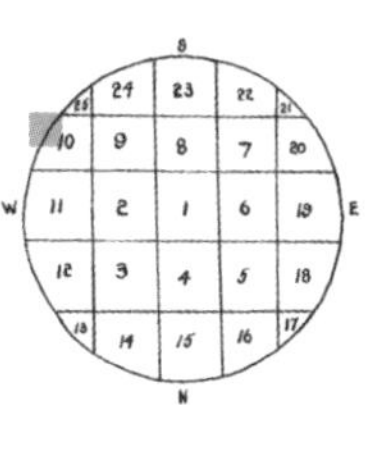

SECTION:
X:I

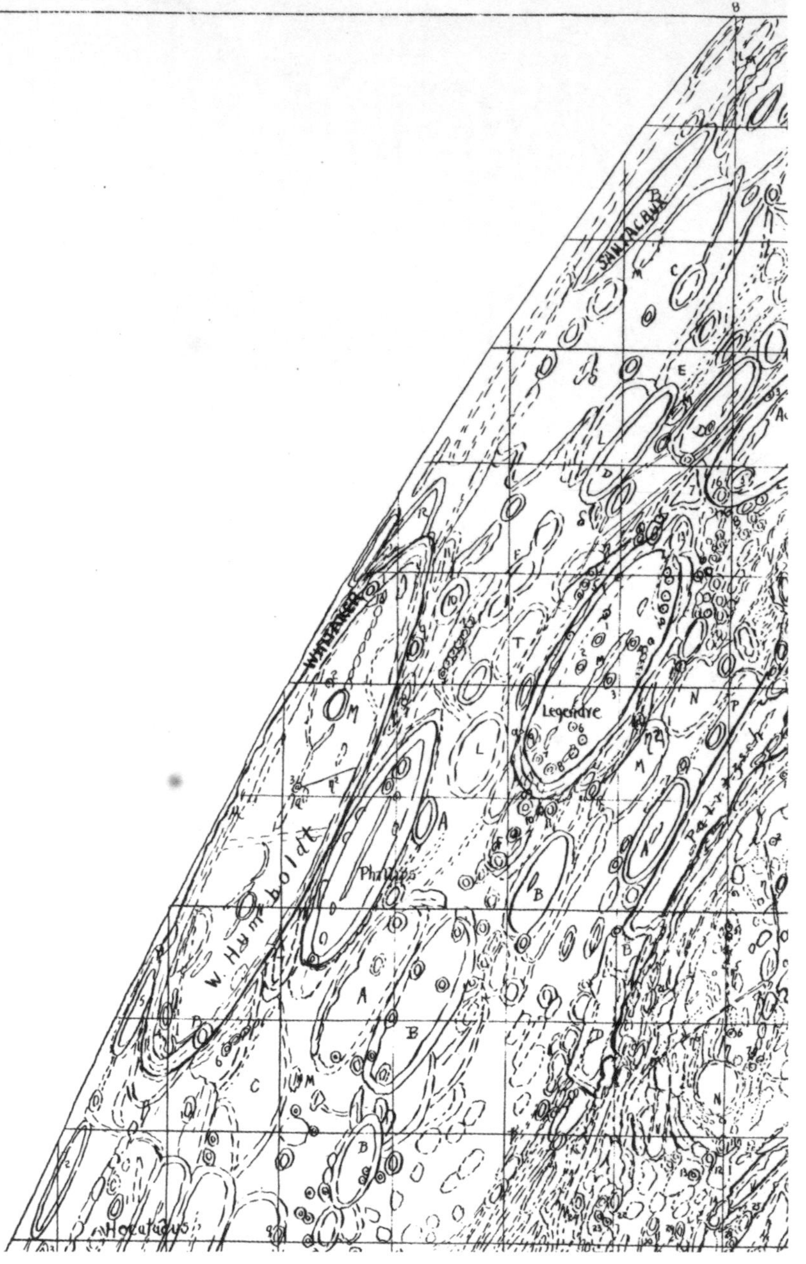

As Director of the Lunar Section of the British Astronomical Association, Wilkins helped to inspire lunar work among astronomers. One of these, **Ewen Whitaker**, went on to become one of the most important figures in 20th century lunar cartography. Wilkins named a crater after him, near the edge of the Moon's visible surface.

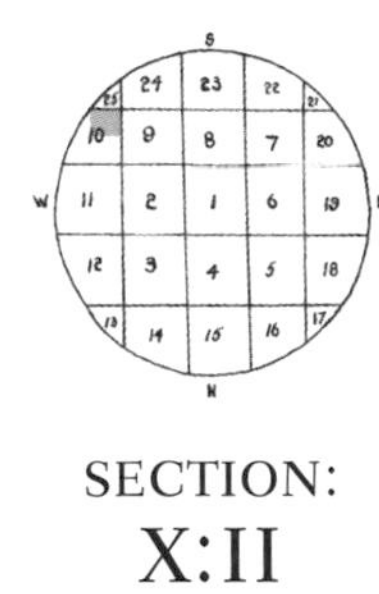
W
E
N
S

SECTION:
X:II

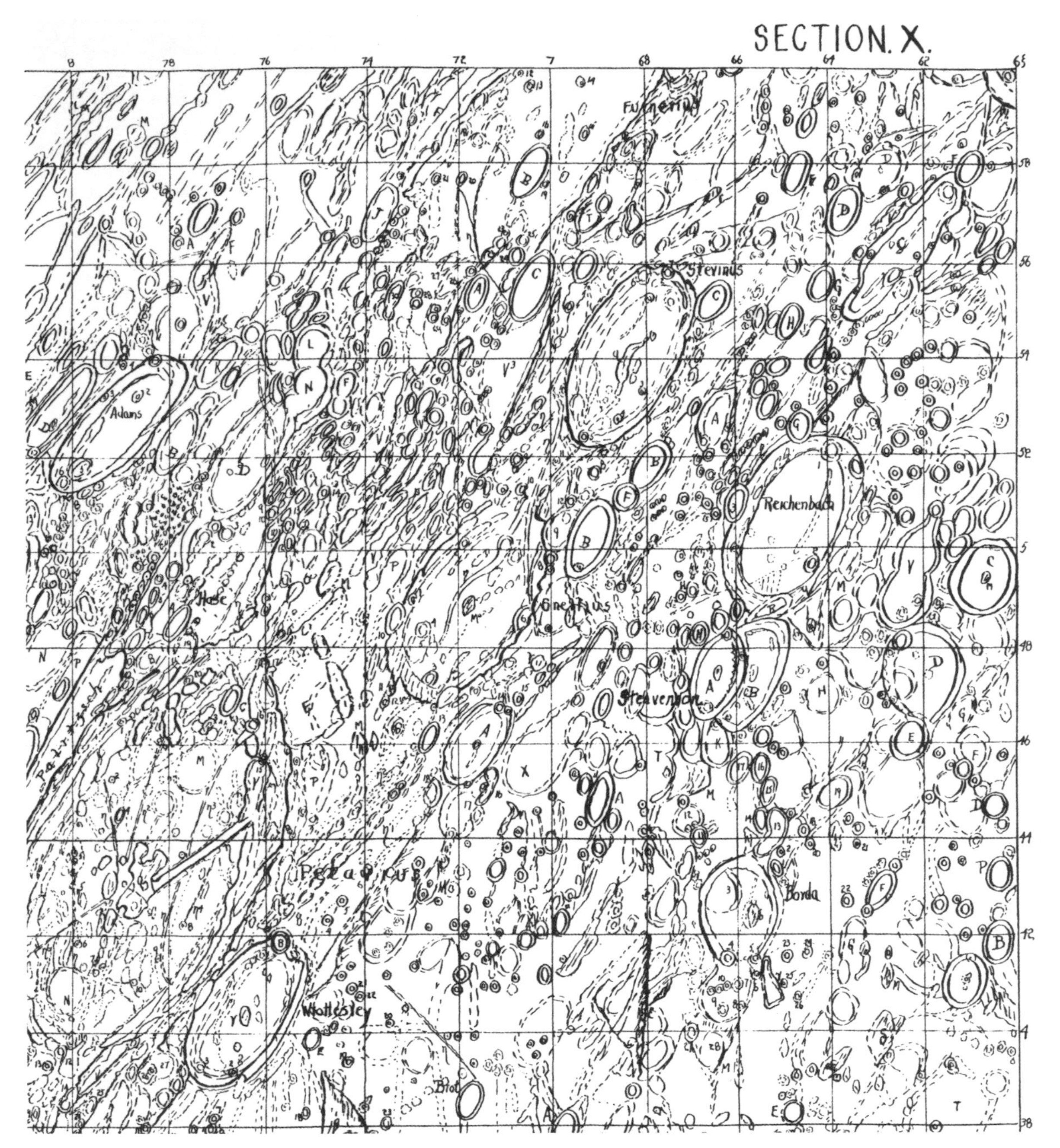
KEY:- Craterlet, Craterpit, Cratercone, Hillock, Clefts, Ridges, Light Streaks, Dark Variable Spots.
SECTION. X.
Furnerius
Stevinus
Adams
Reichenbach
Hase
Snellius
Stevenson
Petavius
Borda
Wottesley
Biot

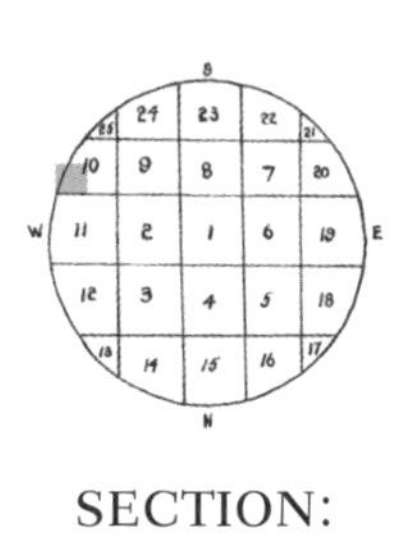

SECTION:
X:III

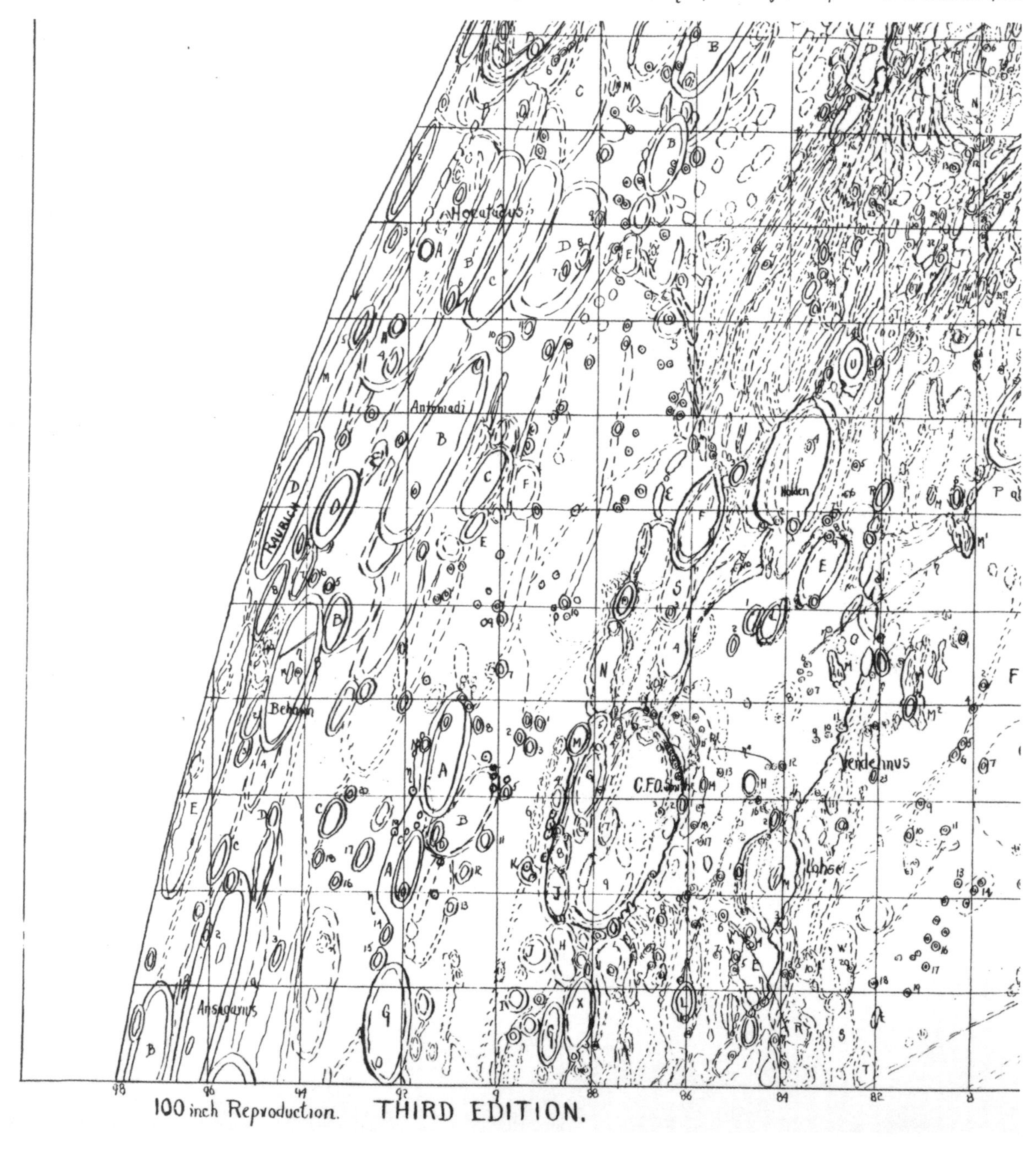

SECTION: X:IV

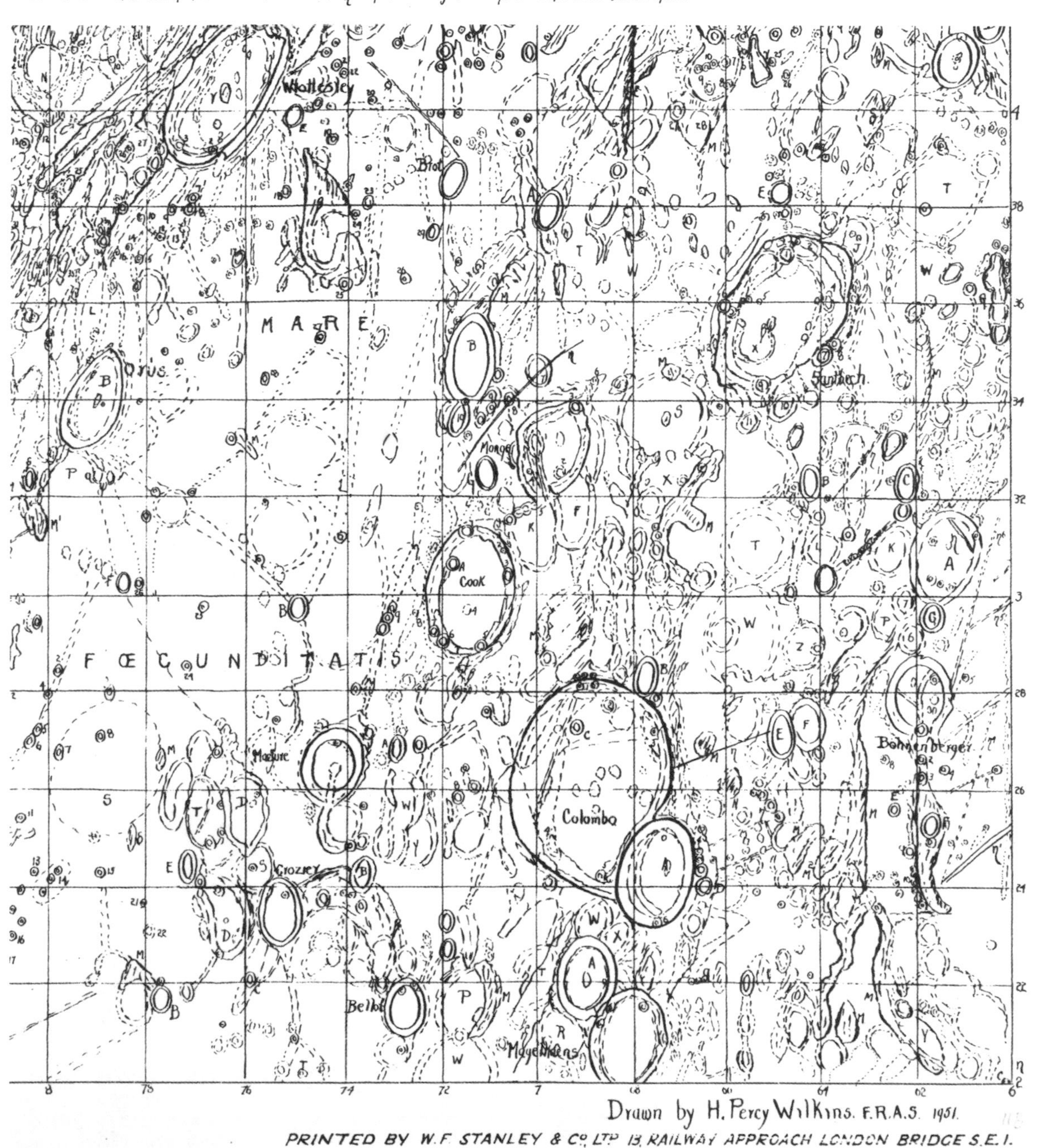
KEY:- ◎ Craterlet, ⊙ Craterpit, ⊕ Cratercone, o Hillock, Clefts, ═ Ridges, ····· Light Streaks, Dark Variable Spots.
MARE
FŒCUNDITATIS
Wrottesley
Biot
Cook
Monge
Santbech
Colombo
Bohnenberger
Crozier
Bellot
Magelhaens
Drawn by H. Percy Wilkins. F.R.A.S. 1951.
PRINTED BY W.F. STANLEY & Co. LTD 13, RAILWAY APPROACH LONDON BRIDGE S.E.1.

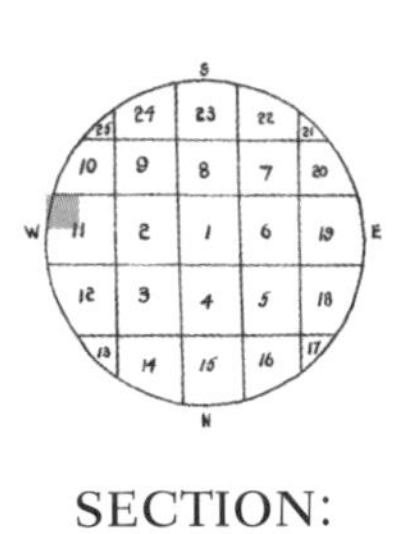

SECTION:
XI:I

KEY:– ◎ Craterlet, ⊙ Craterpit, ⊕ Cratercone, o Hillock, Clefts, ═ Ridges, ······ Light Streaks, Dark Variable Spots.

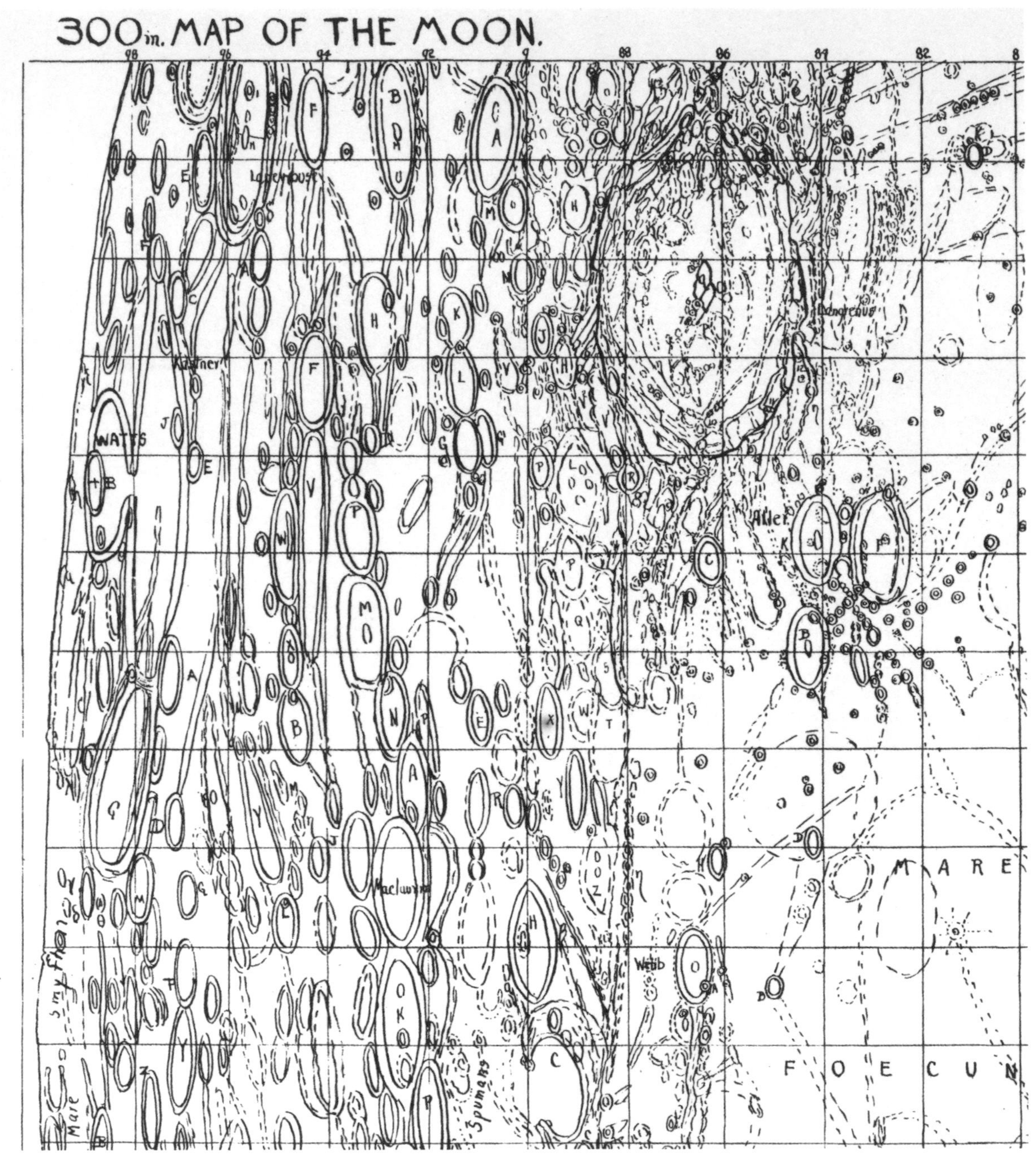

William Henry Smyth (1788–1865) was a Royal Naval officer, hydrographic surveyor and astronomer. **Mare Smythii**, or Smyth's Sea, was first so-named by Smyth's friend, the astronomer John Lee, in 1863.

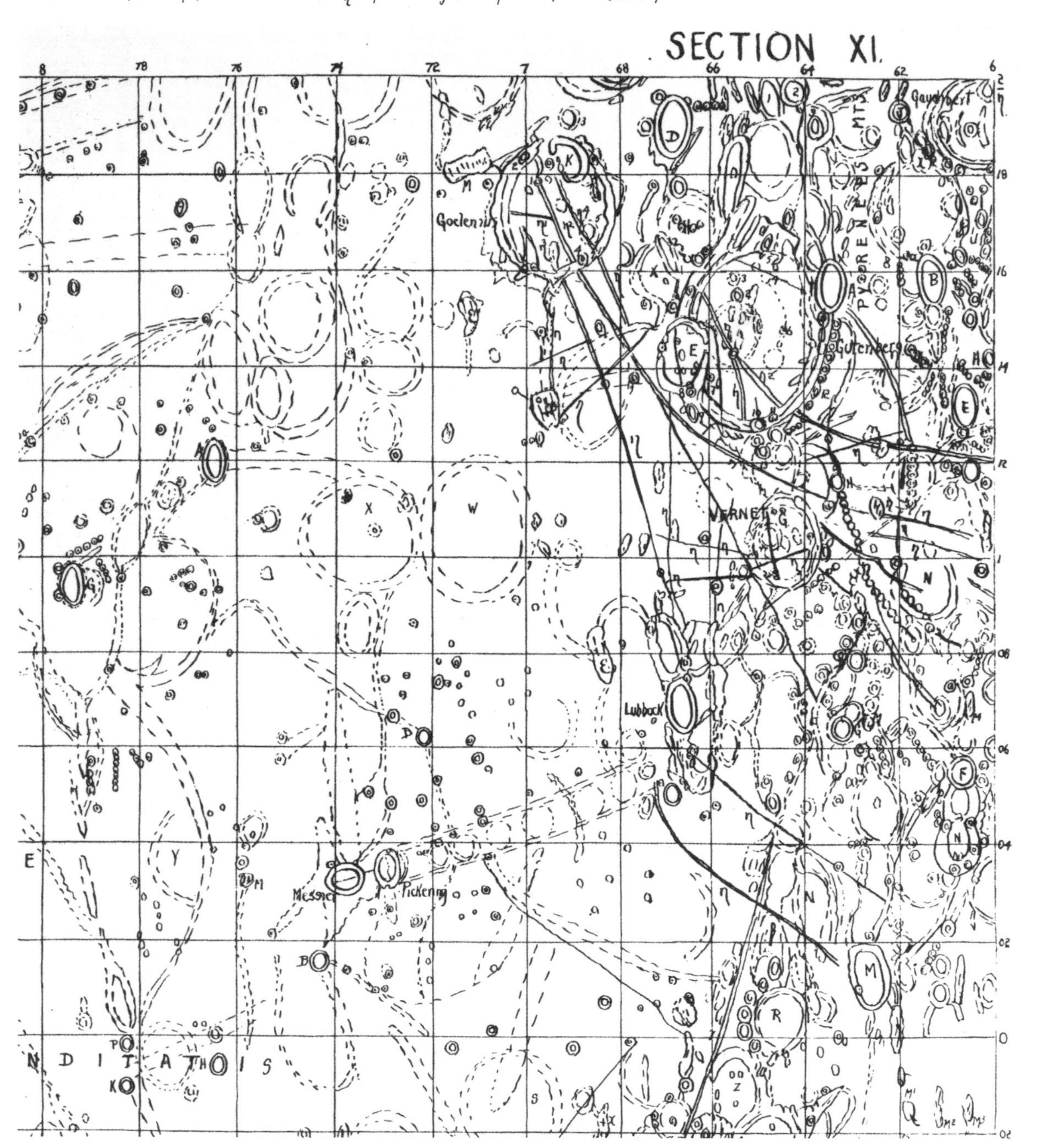

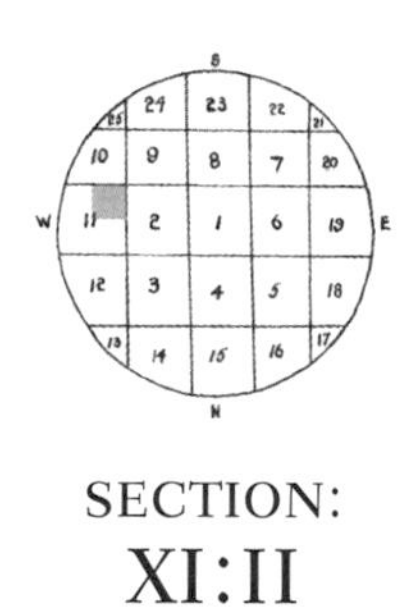

SECTION:
XI:II

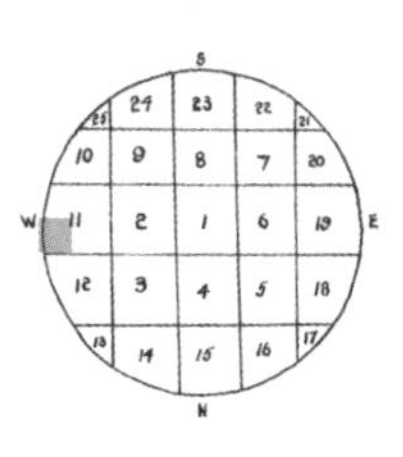

SECTION:
XI:III

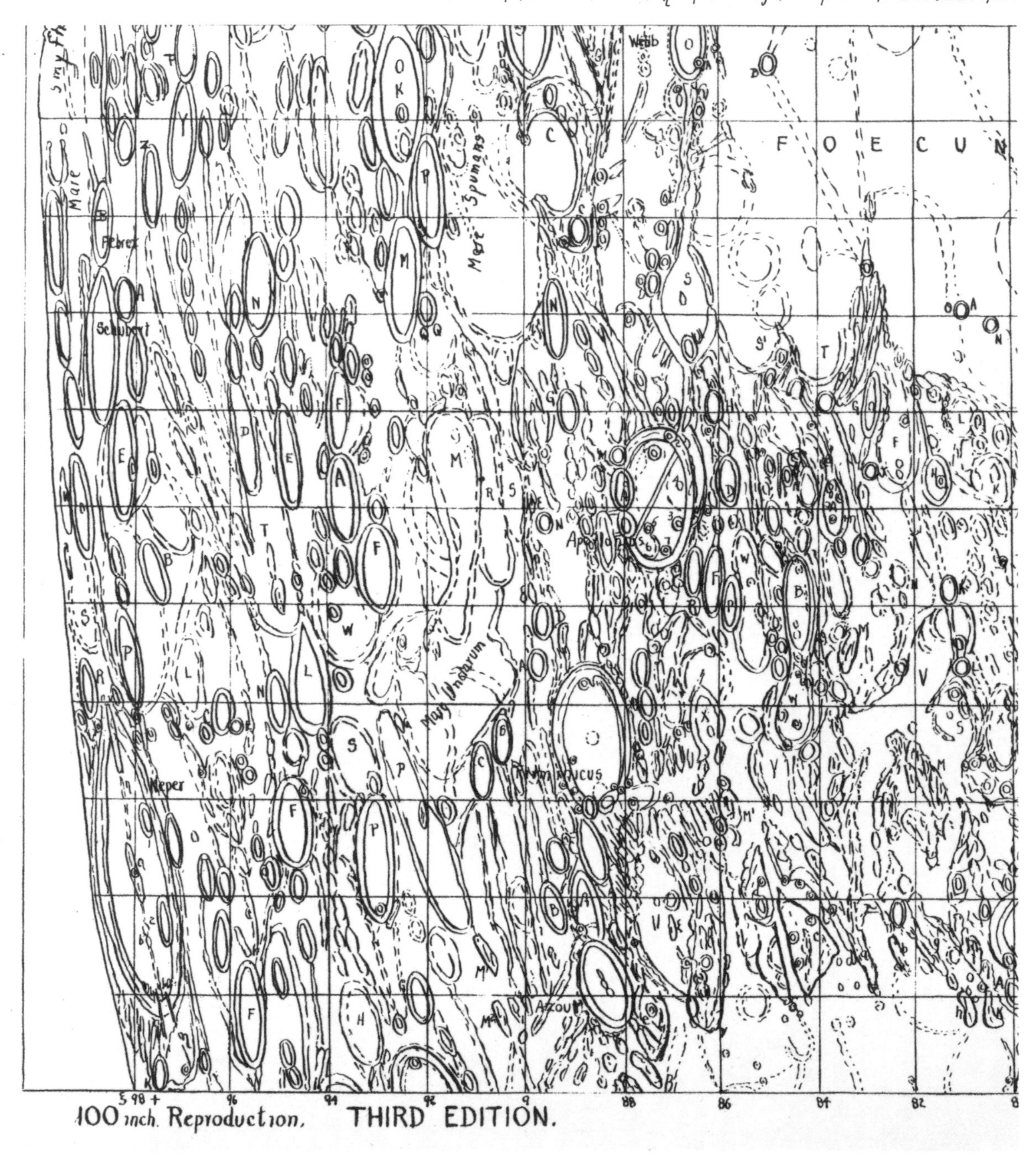
KEY:- Craterlet, Craterpit, Cratercone, Hillock, Clefts, Ridges, Light Streaks, Dark Variable Spots.
F O E C U N
Mare Spumans
Mare Febres
Schubert
Webb
Apollonius
Mare Undarum
Neper
Azout
100 inch. Reproduction. THIRD EDITION.

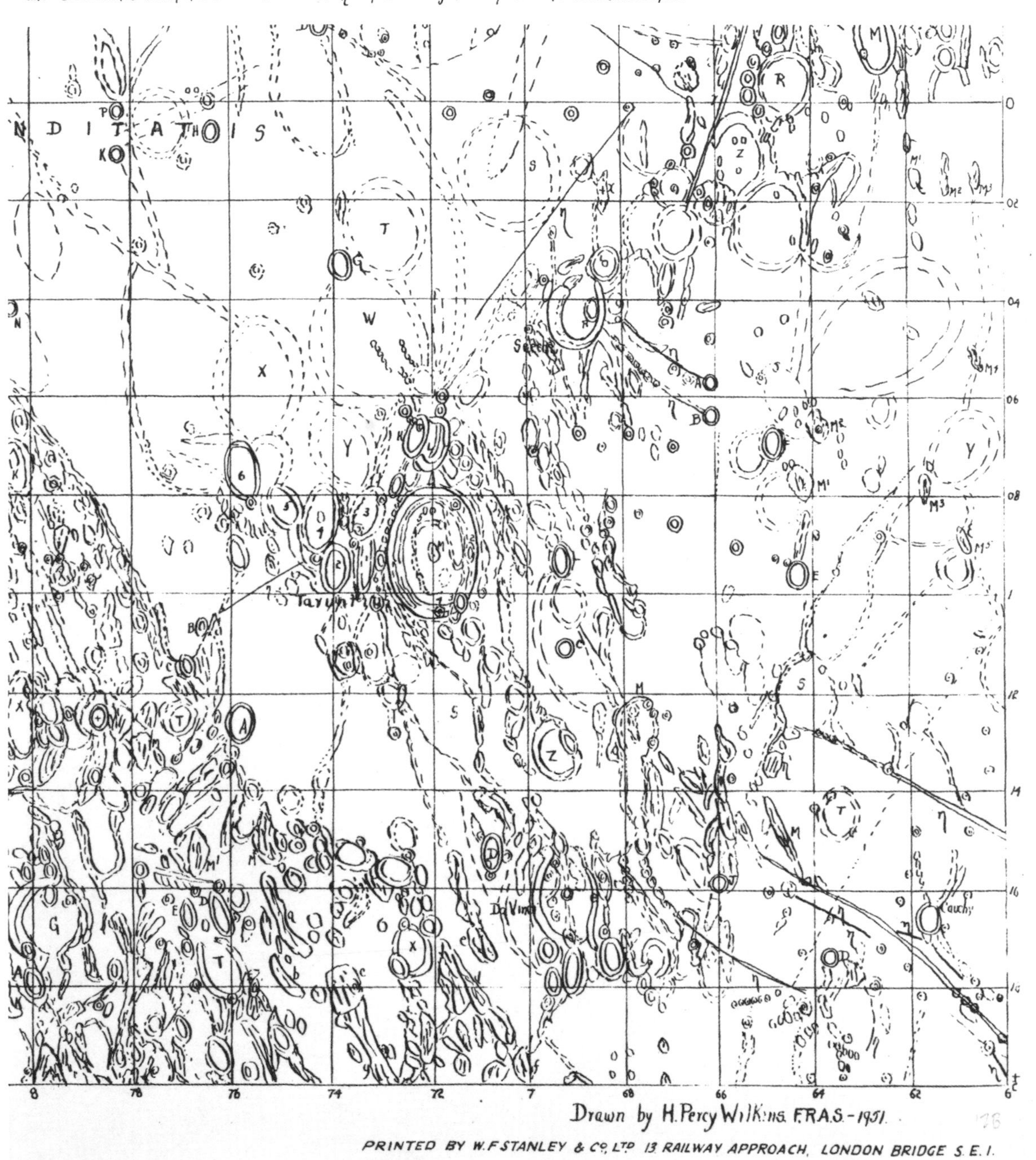
KEY:- Craterlet, Craterpit, Cratercone, Hillock, Clefts, Ridges, Light Streaks, Dark Variable Spots.
Drawn by H. Percy Wilkins F.R.A.S. - 1951.
PRINTED BY W. F. STANLEY & Co. LTD 13 RAILWAY APPROACH, LONDON BRIDGE S.E.1.

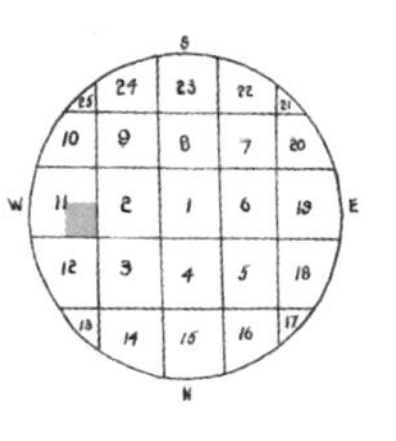

SECTION:
XI:IV

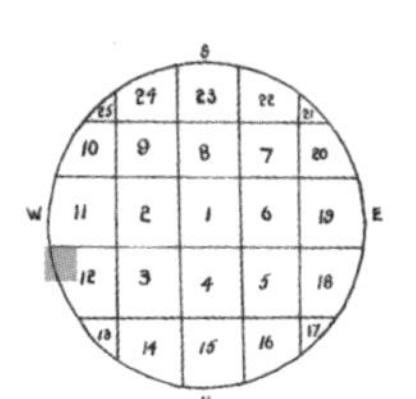

SECTION:
XII:I

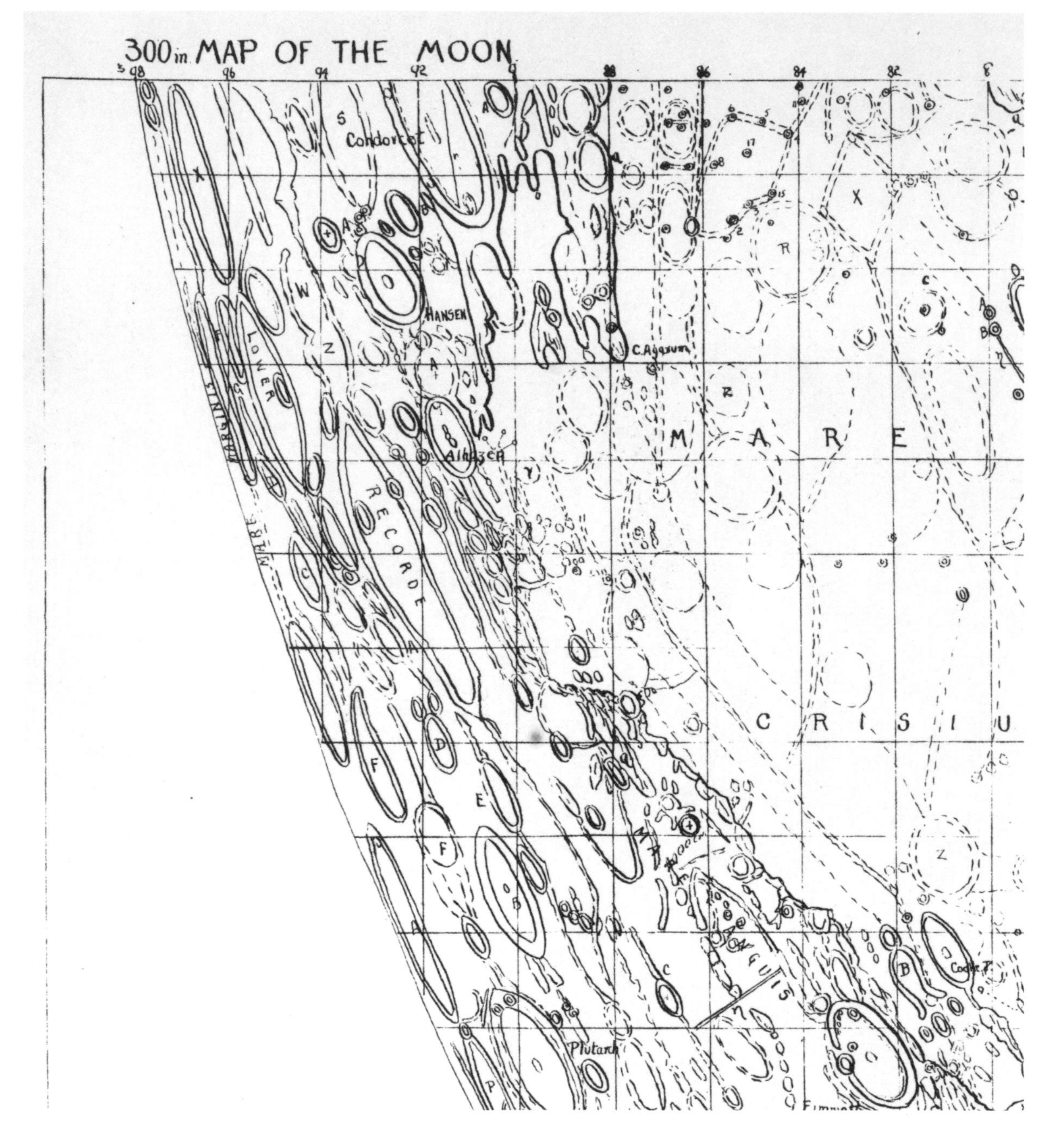

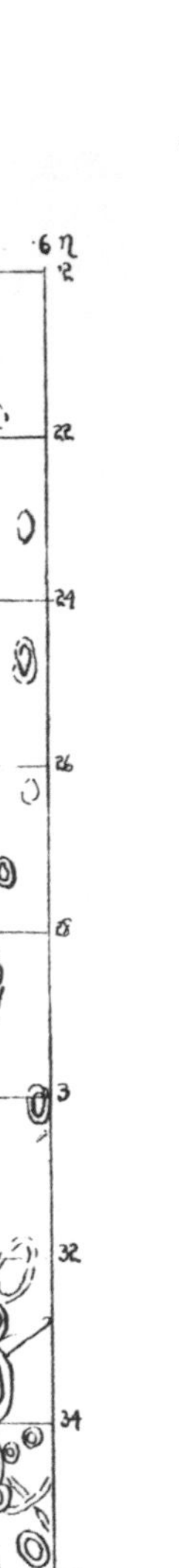

SECTION:
XII:II

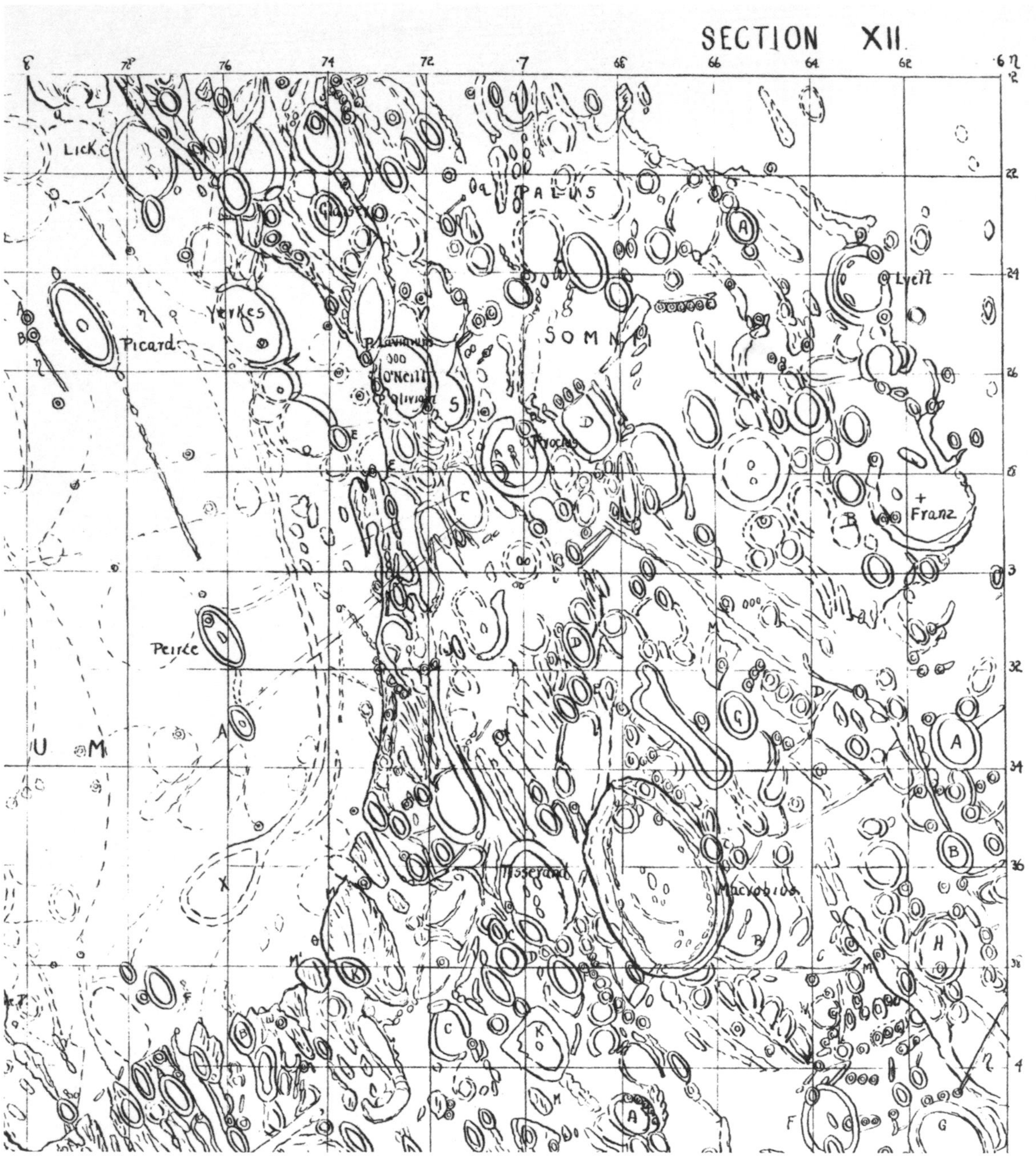

This sheet includes the location of the feature (not visible on this map) which so damaged Wilkins's reputation. At one edge of the crater **O'Neill** (another of Wilkins's unofficial names), he publicly confirmed the existence of a natural bridge on the Moon. It's now understood to be an illusion, caused by the way shadows fall in this particular area.

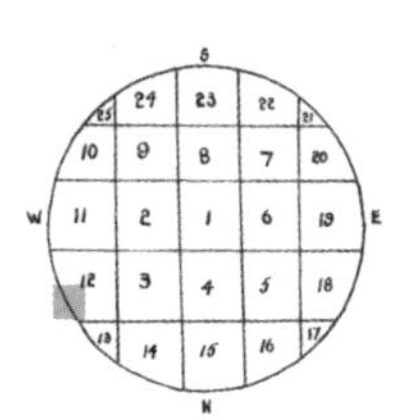

SECTION:
XII:III

KEY:- ◎ Craterlet, ⊙ Craterpit, ⊕ Cratercone, ○ Hillock, ─η─ Clefts, ═══ Ridges, ······ Light Streaks, ▦ Dark Variable Spots.

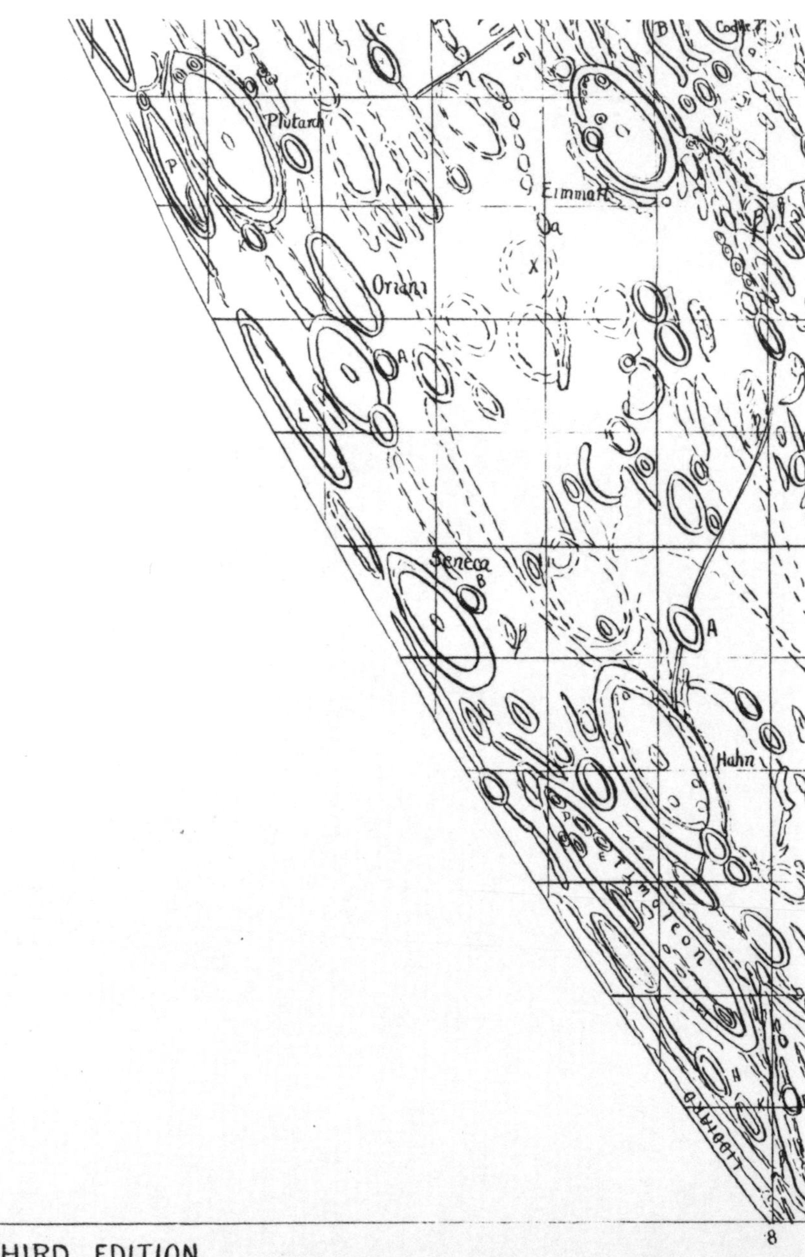

100 in. Reproduction THIRD EDITION.

KEY:- ◎ Craterlet, ⊙ Craterpit, ⊕ Cratercone, o Hillock, Clefts, ═ Ridges, ······ Light Streaks, Dark Variable Spots.

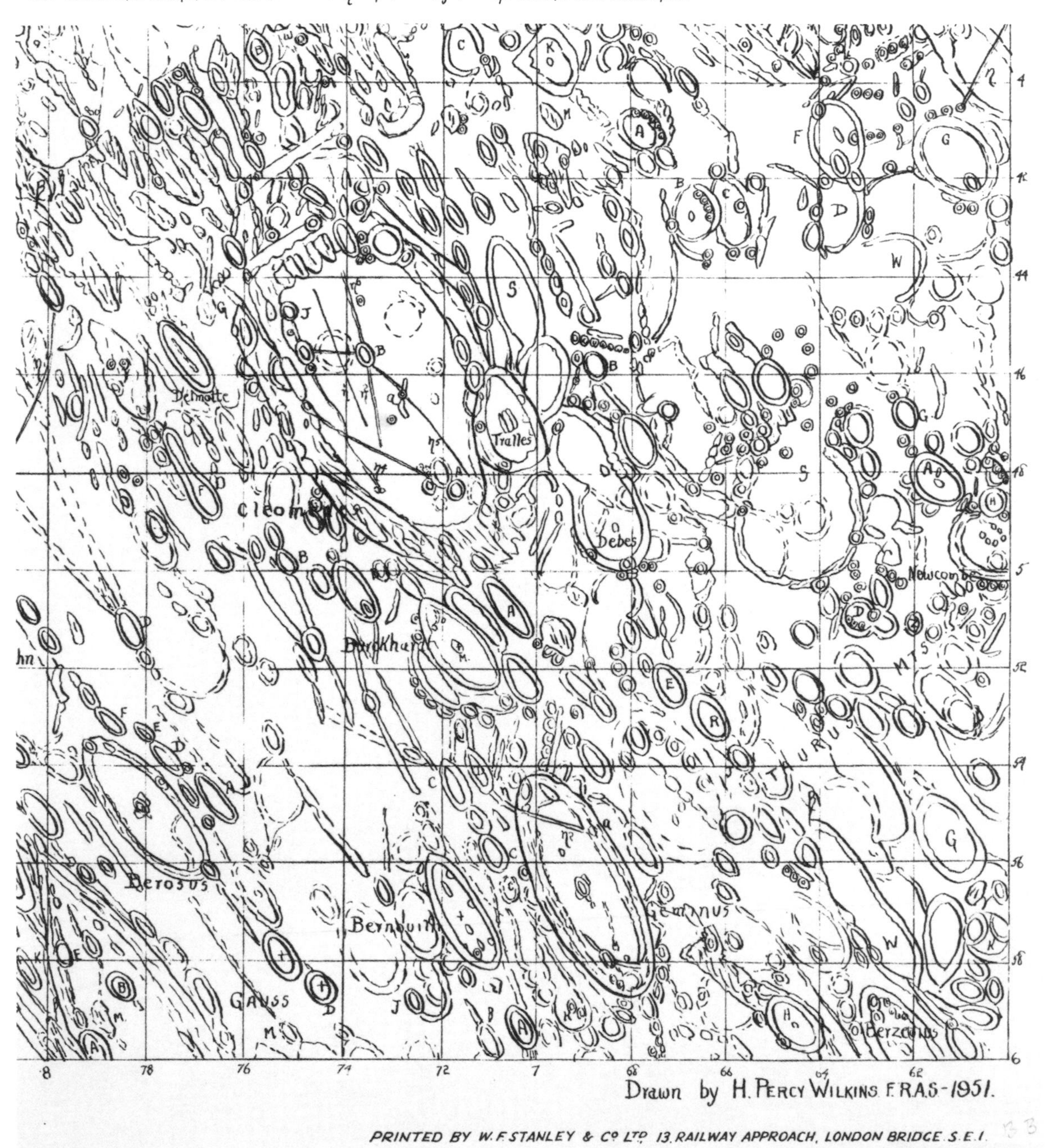

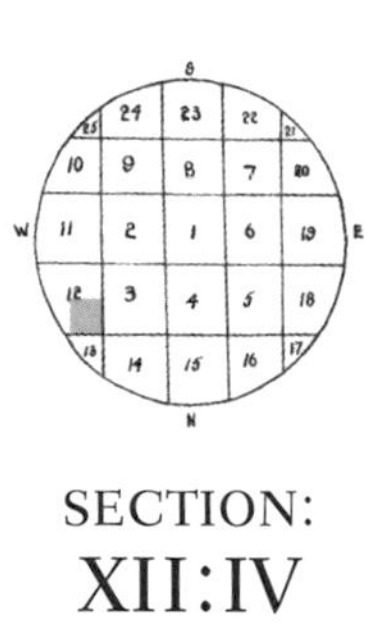

SECTION:
XII:IV

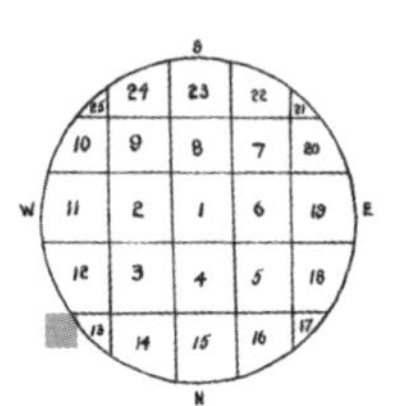

SECTION:
XIII:I

KEY:- Craterlet, Craterpit, Cratercone, Hillock, Clefts, Ridges, Light Streaks, Dark Variable Spots.

300 in. MAP OF THE MOON.

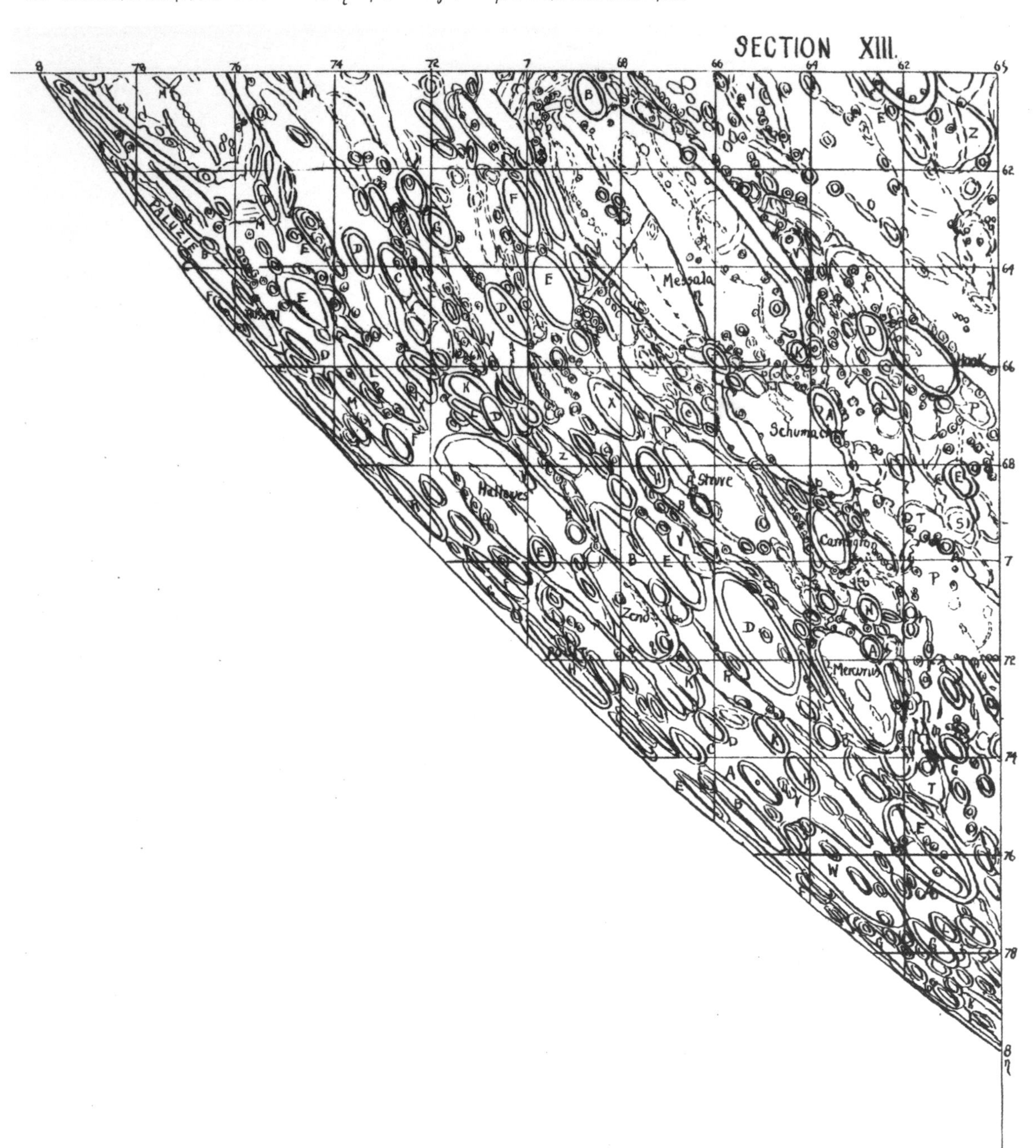

SECTION:
XIII:II

Messala is named after Māshā'allāh ibn Atharī, a Jewish astrologer and mathematician from Basra. He wrote works on the movement of the spheres and on eclipses, and was one of the leading astrologers in 8th–9th century Baghdad.

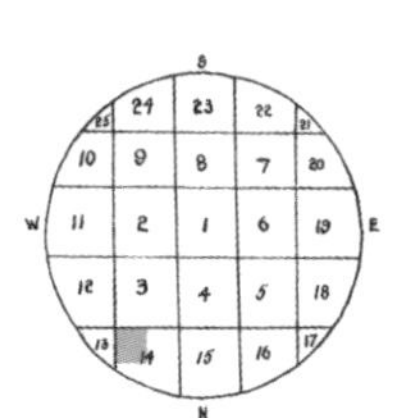

SECTION:
XIV:I

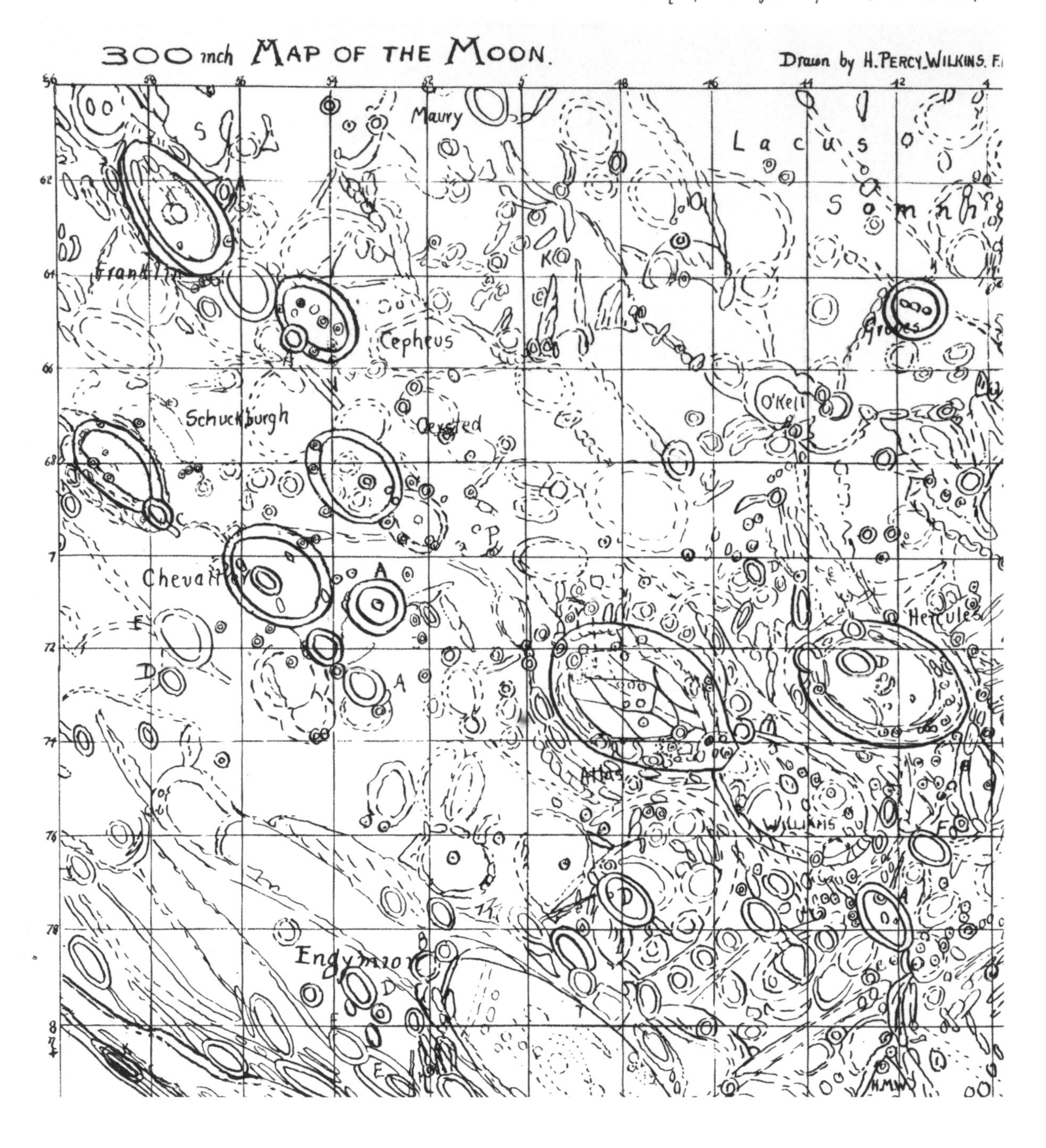
KEY:- Craterlet, Craterpit, Cratercone, Hillock, Clefts, Ridges, Light Streaks, Dark Variable Spots.
300 inch MAP OF THE MOON.
Drawn by H. PERCY WILKINS.
Maury
Lacus
Somnii
Franklin
Cepheus
Schuckburgh
Oersted
O'Kell
Chevallier
Hercules
Atlas
Williams
Endymion

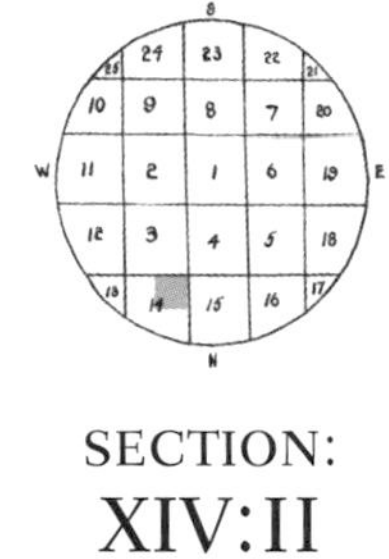

SECTION:
XIV:II

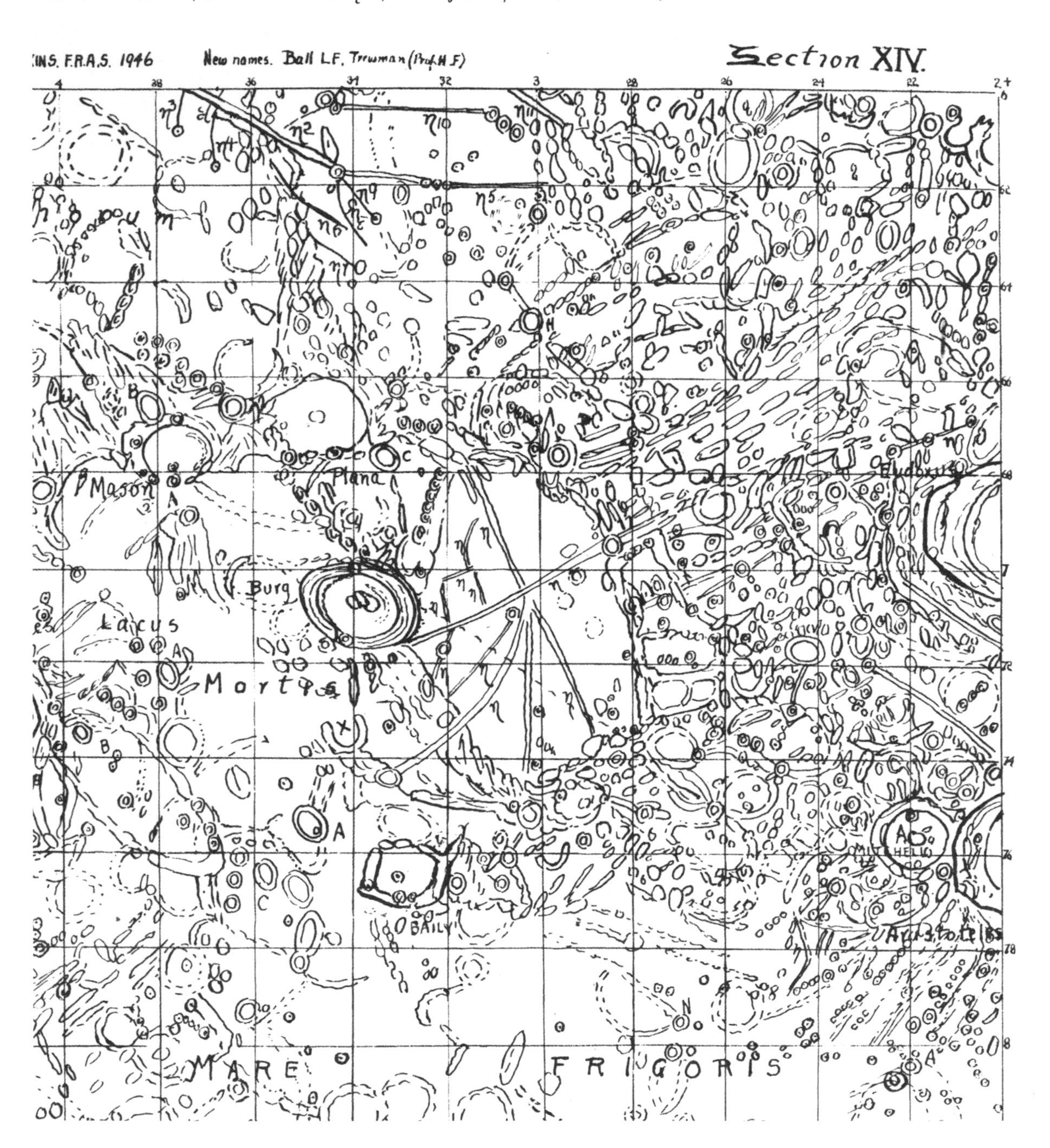
KEY:- ◎ Craterlet, ⊙ Craterpit, ⊕ Cratercone, o Hillock, Clefts, Ridges, Light Streaks, Dark Variable Spots.
INS. F.R.A.S. 1946
New names. Ball L.F. Trouman (Prof. H.F.)
Section XIV.
Mason
Plana
Burg
Lacus
Mortis
BAILY
MITCHELL
Aristoteles
MARE
FRIGORIS

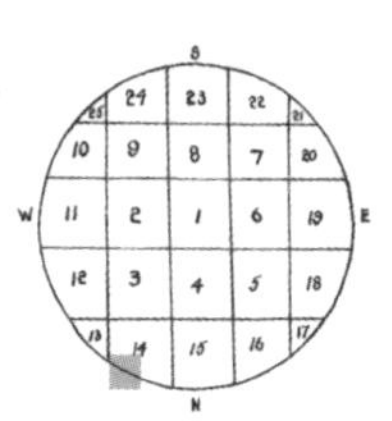

SECTION:
XIV:III

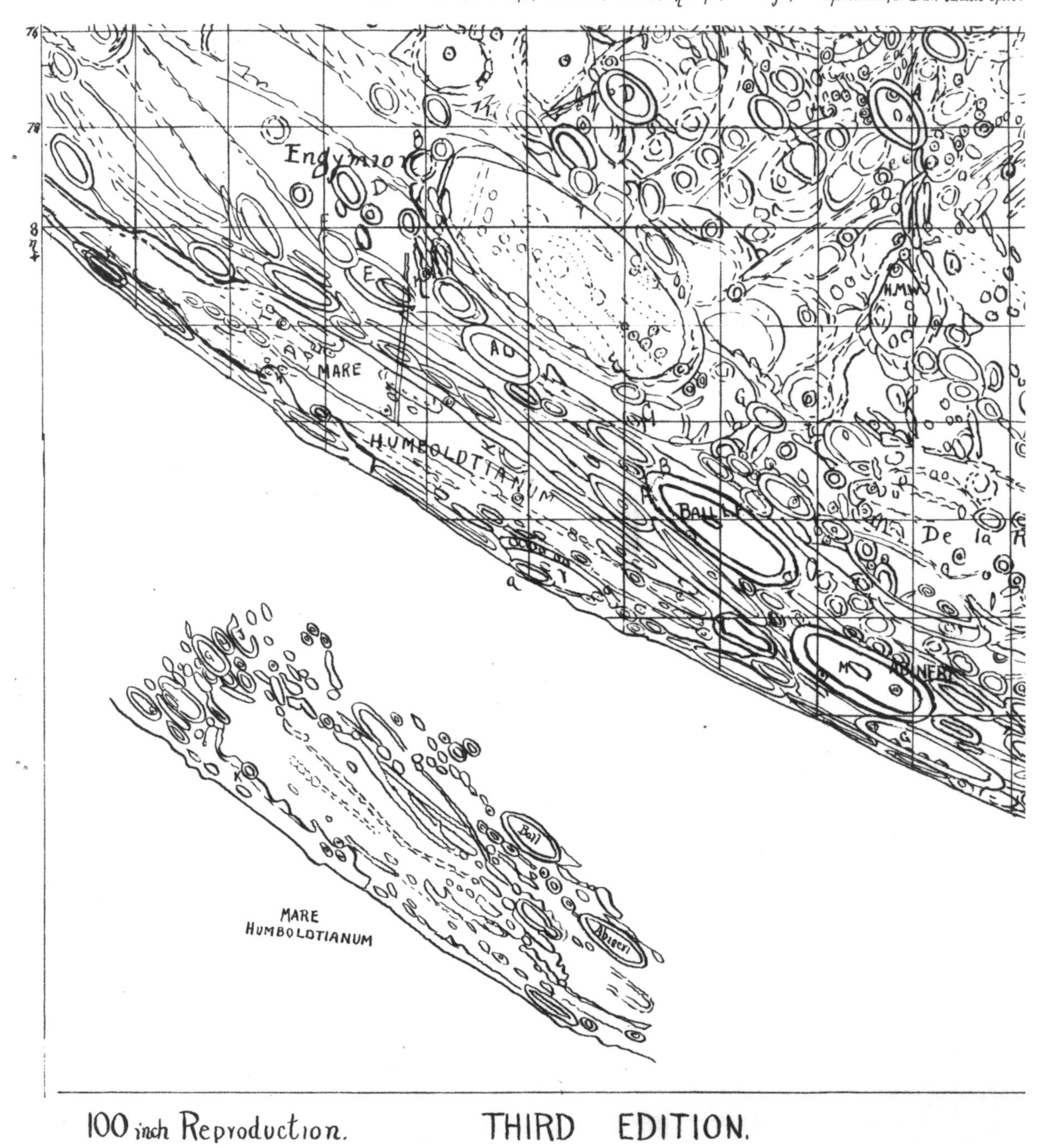

Although only one side of the Moon ever faces Earth, the Moon wobbles slightly. This movement is known as libration. It means that the edges of the visible surface of the Moon change, making certain craters more visible at some times than at others. **Mare Humboldtianum** is one such region, so Wilkins included additional detail alongside his main lunar map.

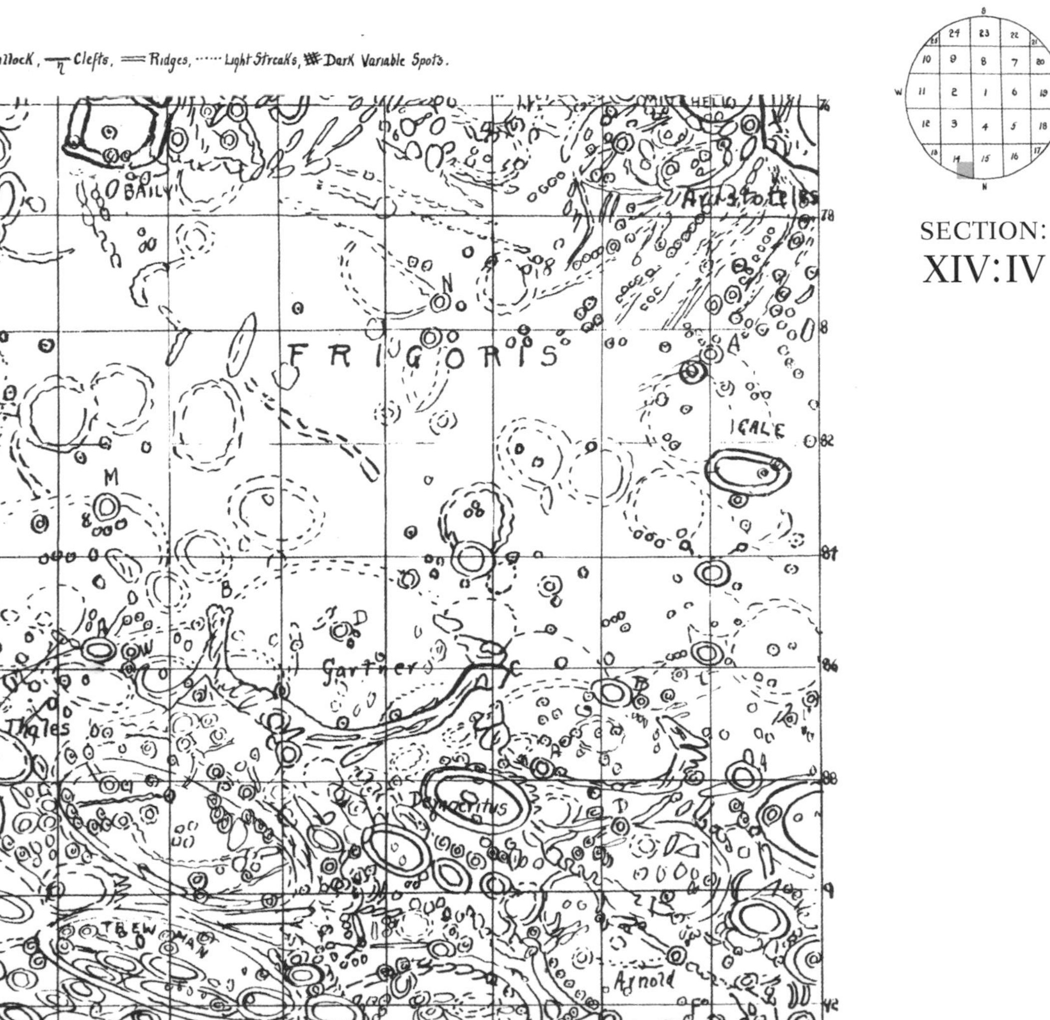

SECTION:
XIV:IV

PRINTED BY W.F. STANLEY & C° L™ 13, RAILWAY APPROACH, LONDON BRIDGE. S.E.1.

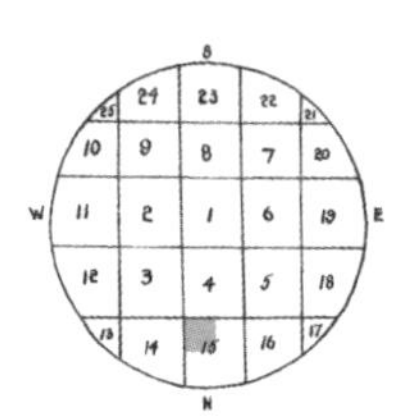

SECTION:
XV:I

KEY:- ◎ Craterlet, ⊙ Craterpit, ⊕ Cratercone, ○ Hillock, Clefts, Ridges, ······ Light Streaks, Dark Variable Spots.

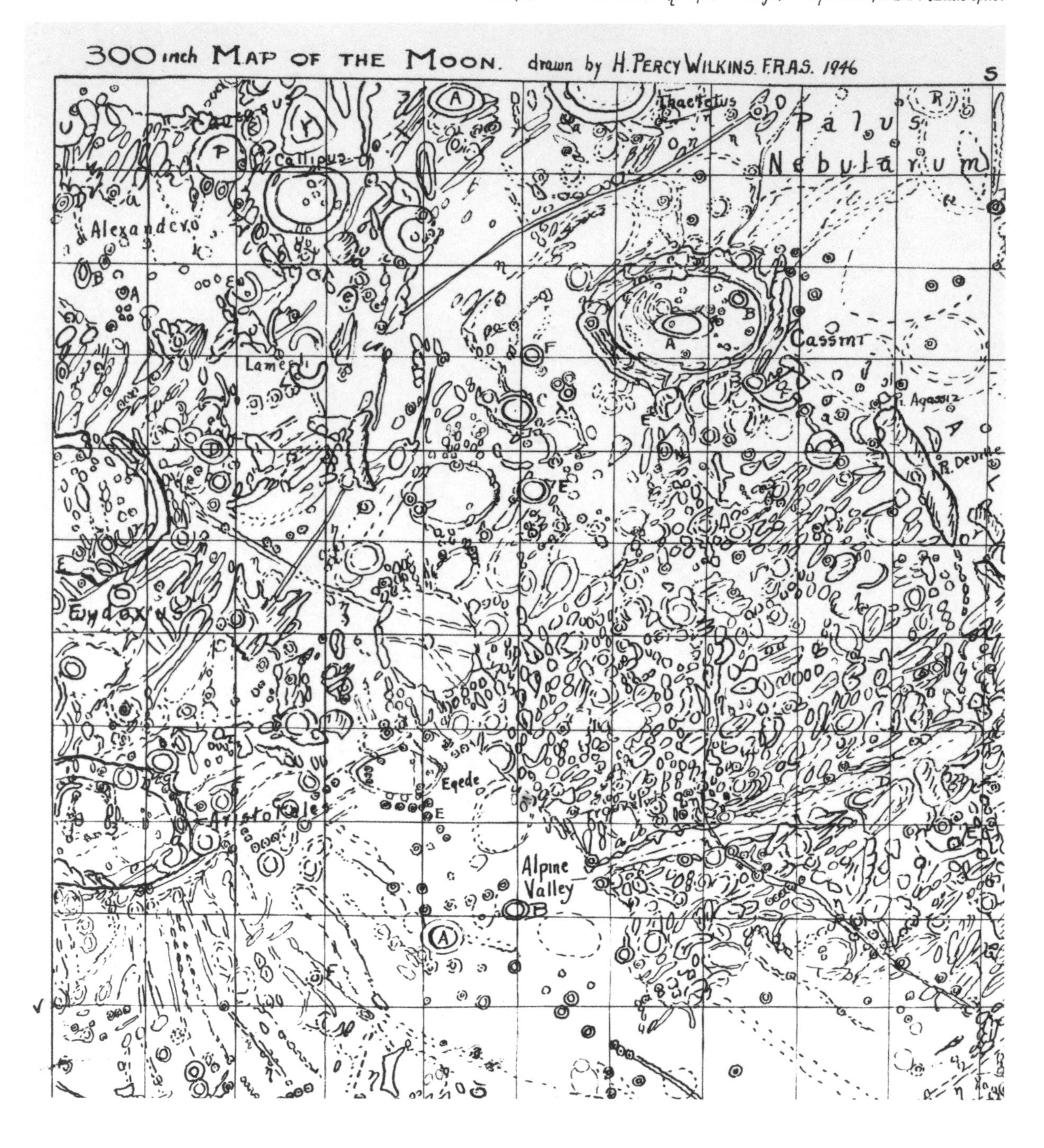

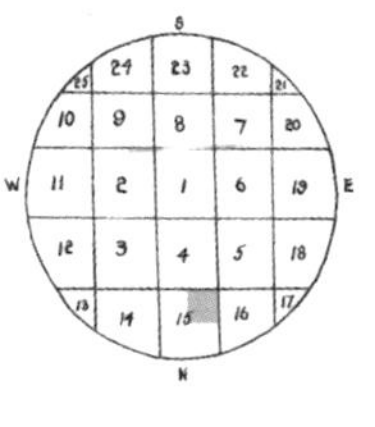

SECTION:
XV:II

KEY:- ◎ Craterlet, ⊙ Craterpit, ⊕ Cratercone, o Hillock, Clefts, ═ Ridges, ······ Light Streaks, Dark Variable Spots.

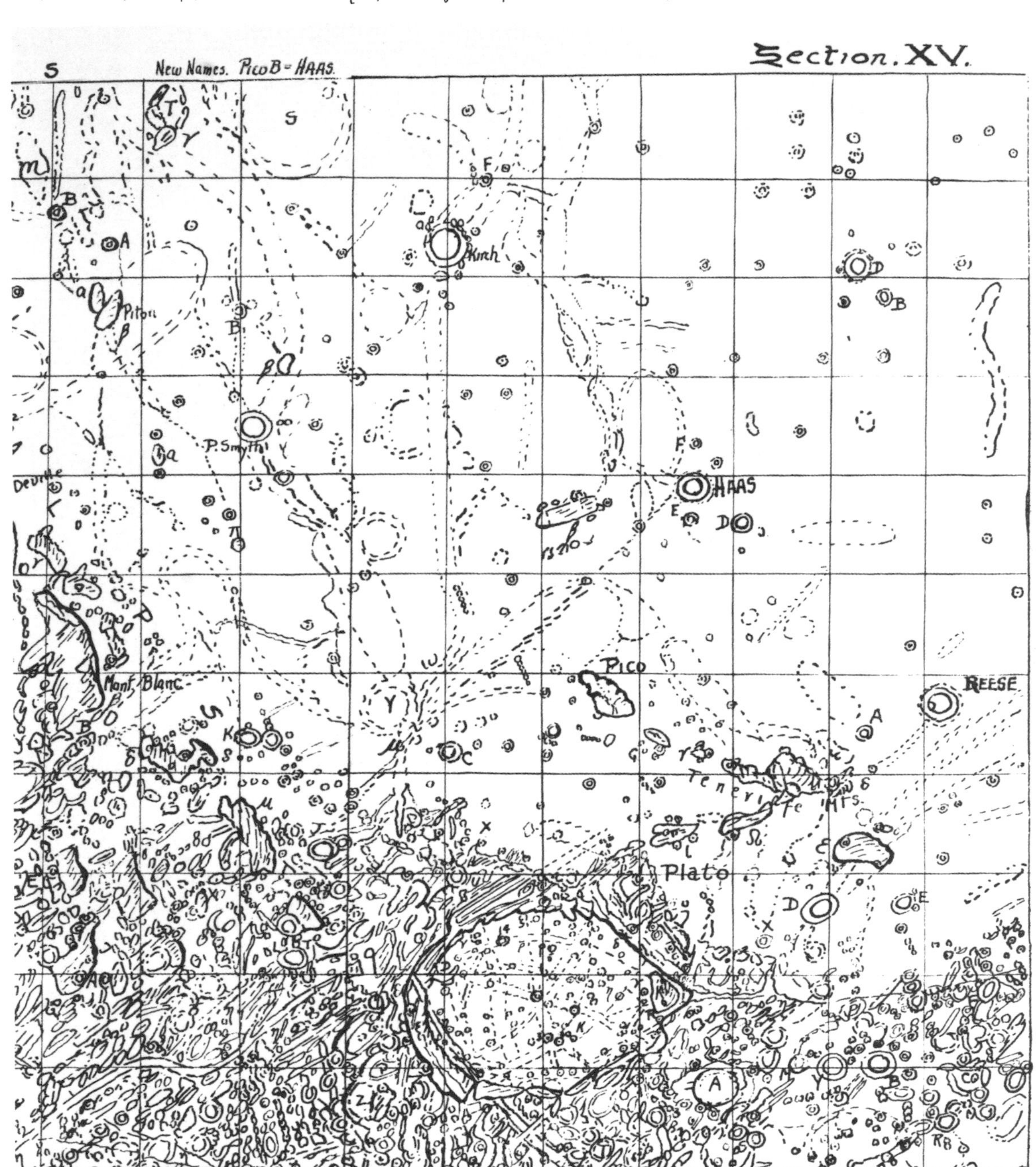

Mons Pico is one of the most impressive lunar mountains to observe. Because it is isolated on the Mare Imbrium (the 'Sea of Rains'), it stands out very sharply in low light, casting a long shadow over the smooth lava plain.

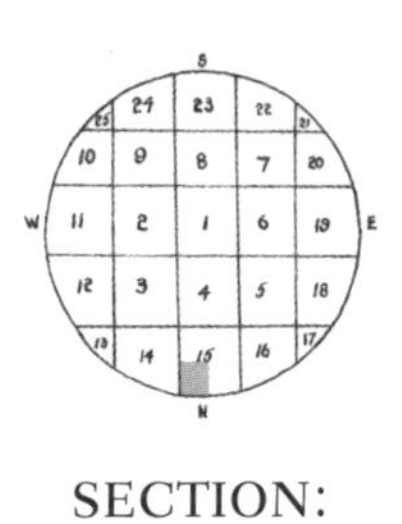

SECTION:
XV:III

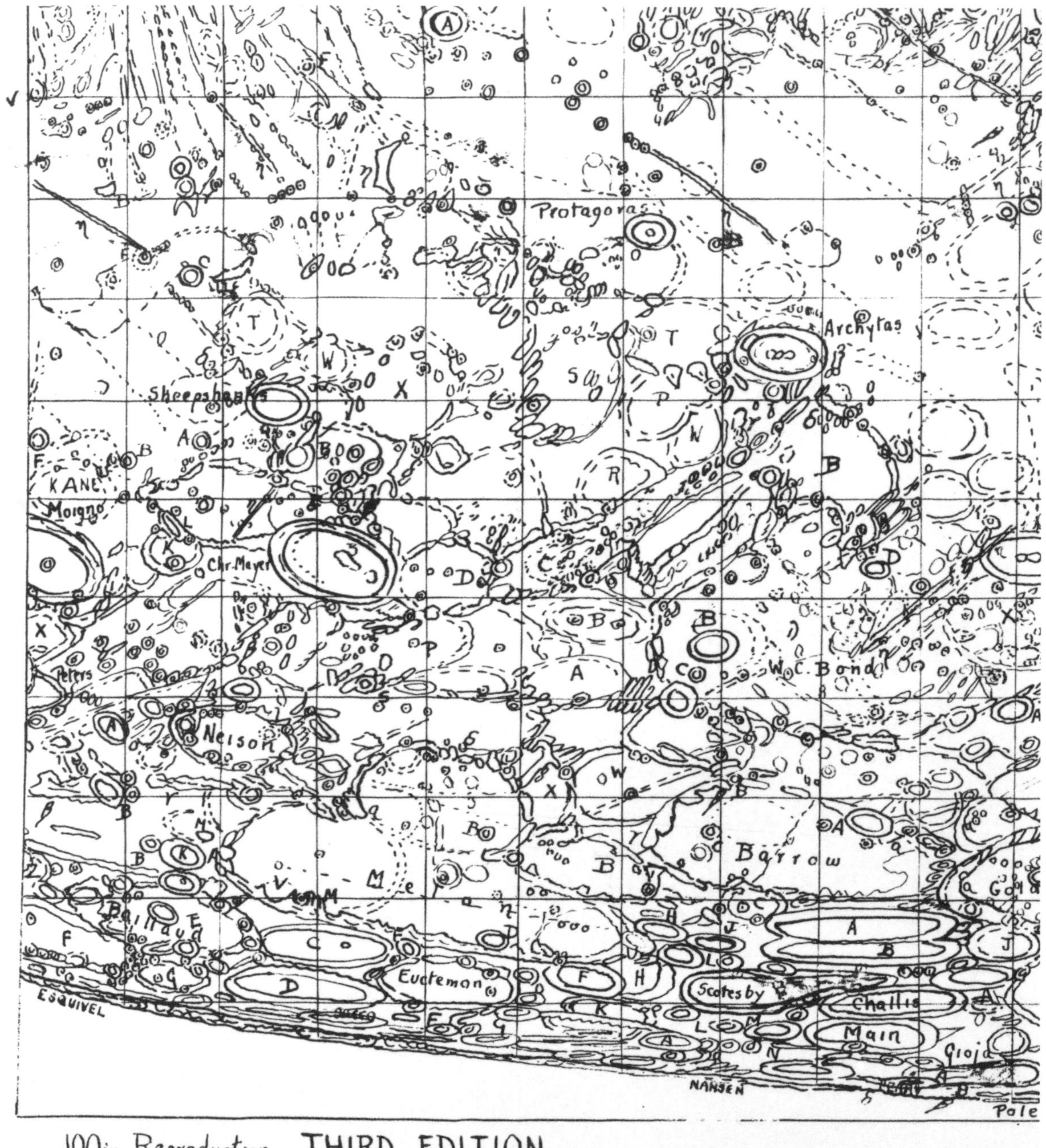

100 in. Reproduction. THIRD EDITION.

Key:- Craterlet, Craterpit, Cratercone, Hillock, Clefts, Ridges, ... Light Streaks, Dark Variable Spots.

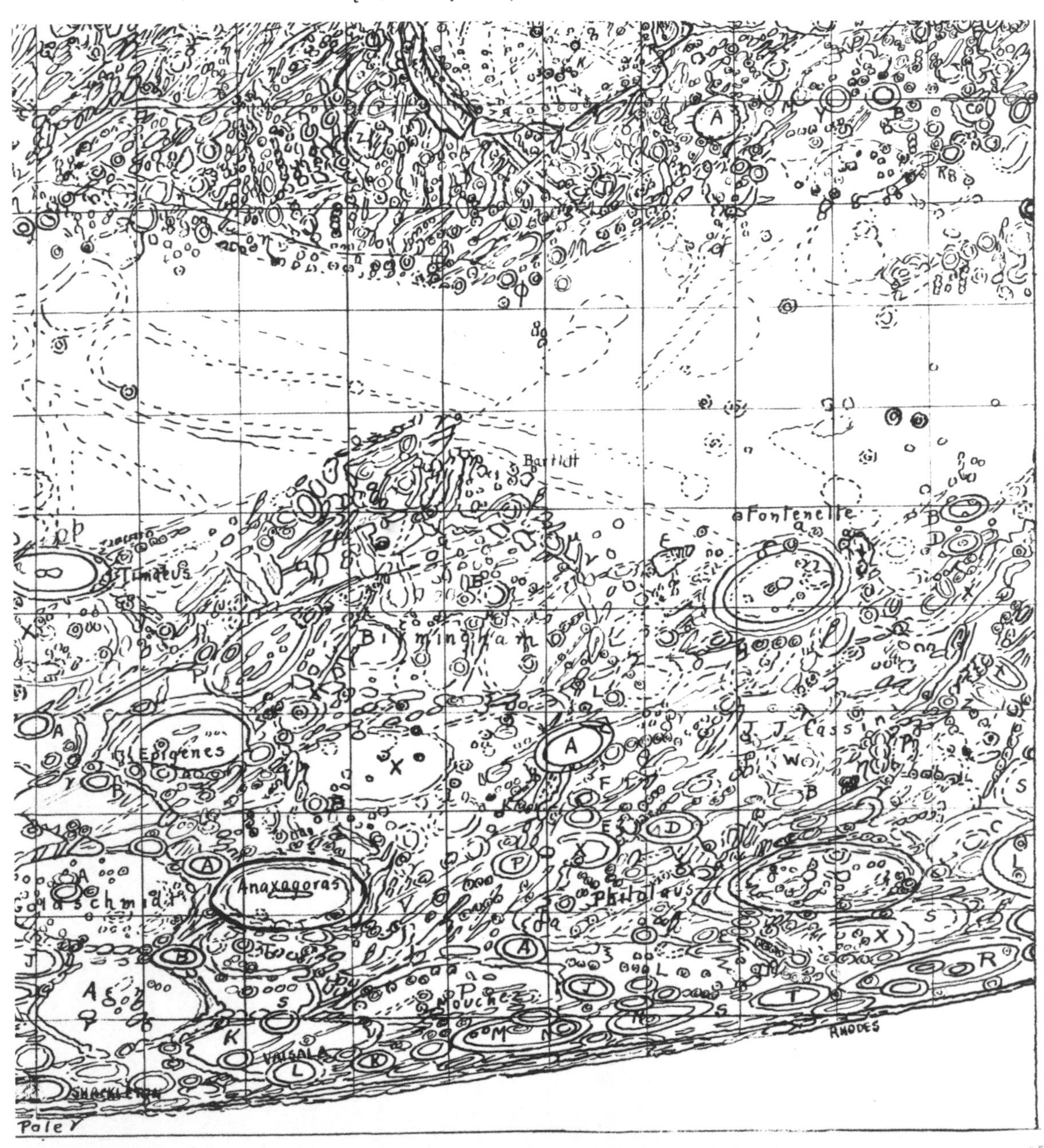

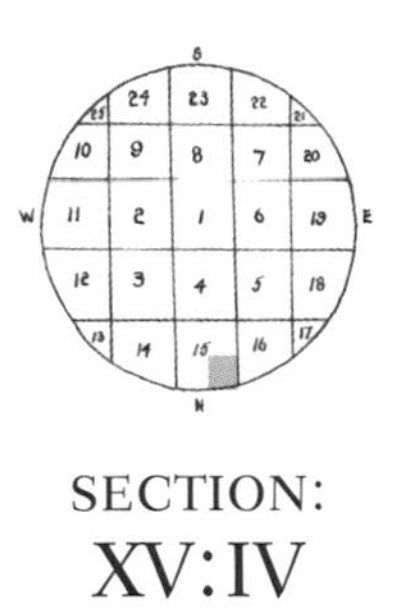

SECTION:
XV:IV

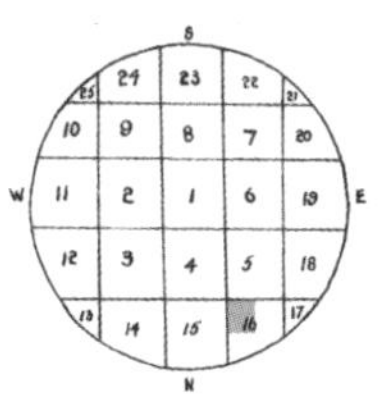

SECTION:
XVI:I

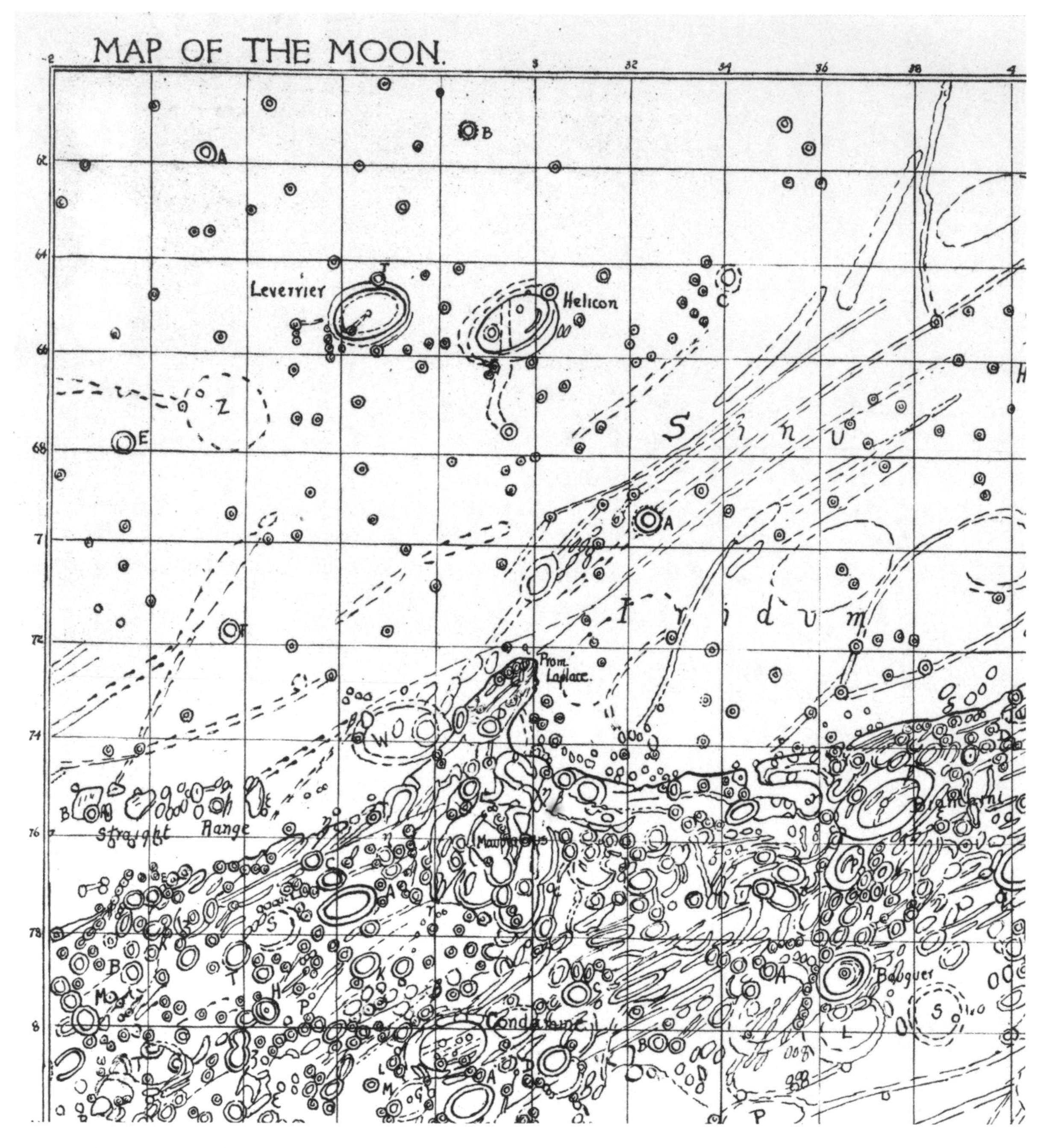

Sinus Iridum (the 'Bay of Rainbows'), is an ancient crater which has been flooded with lava, which also covered one part of the crater wall. This gives it the scoop-shaped appearance of a bay at the edge of Mare Imbrium.

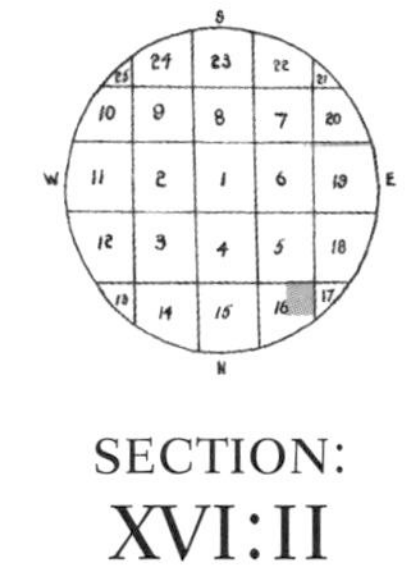

SECTION: XVI:II

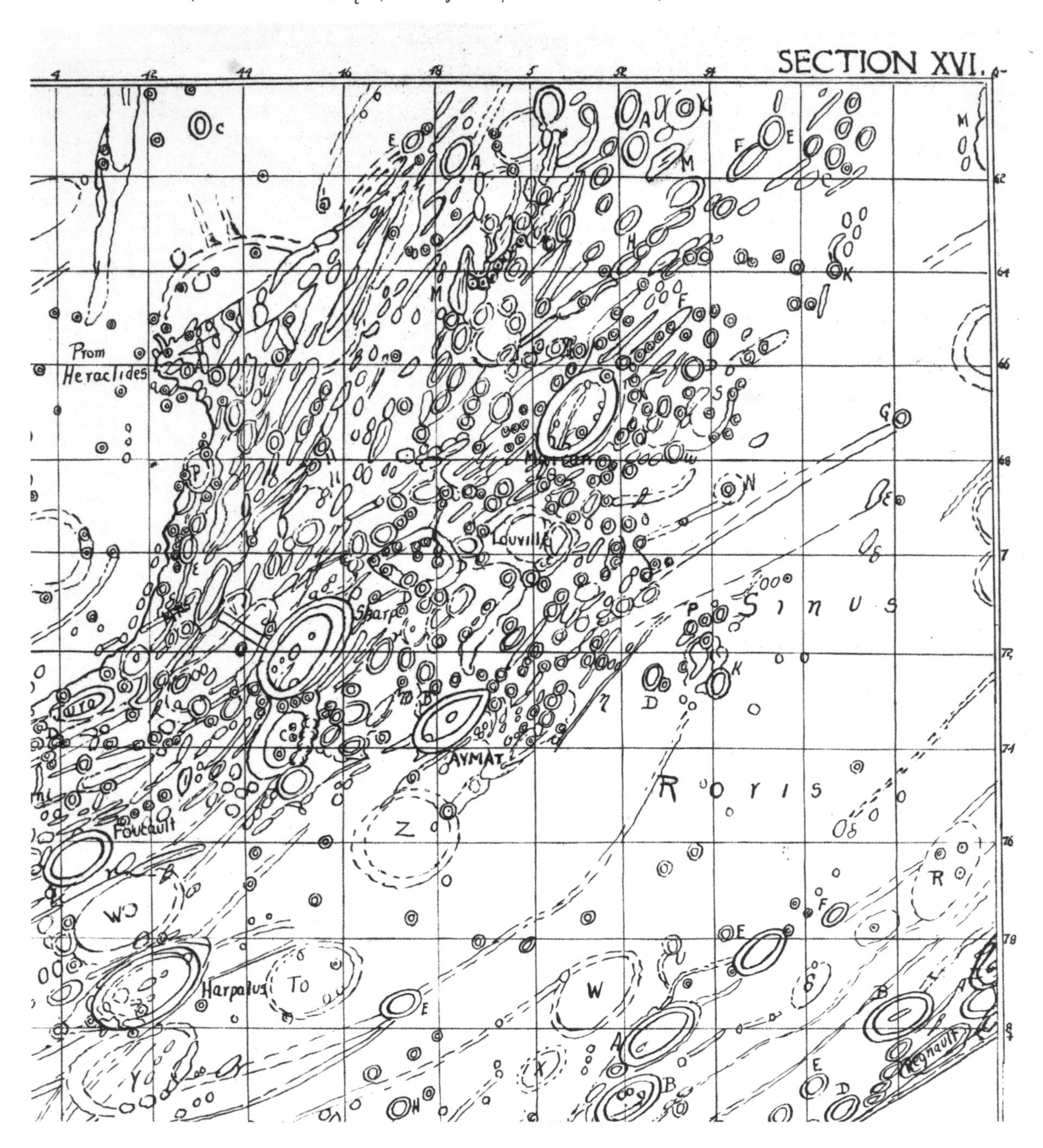

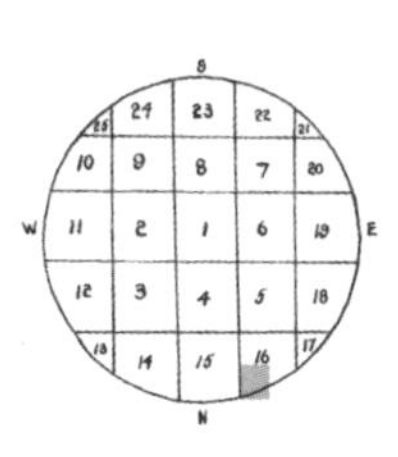

SECTION:
XVI:III

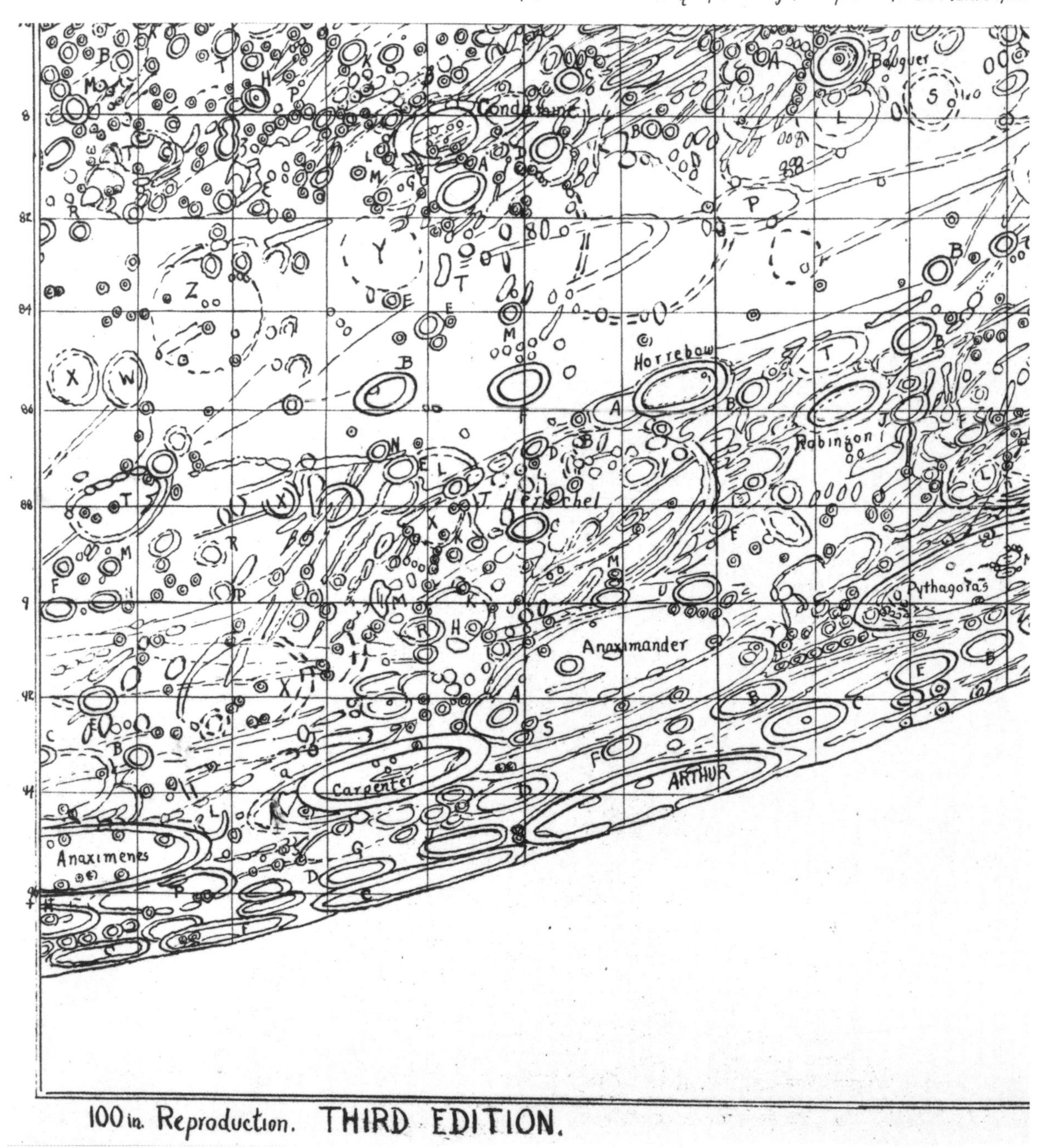

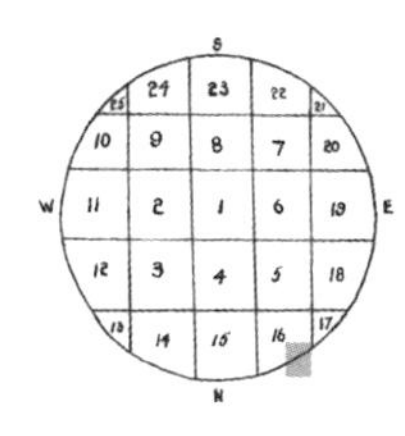

SECTION:
XVI:IV

KEY:- ◎ Craterlet, ⊙ Craterpit, ⊕ Cratercone, ○ Hillock, ‑η‑ Clefts, ═ Ridges, ······ Light Streaks, ▦ Dark Variable Spots.

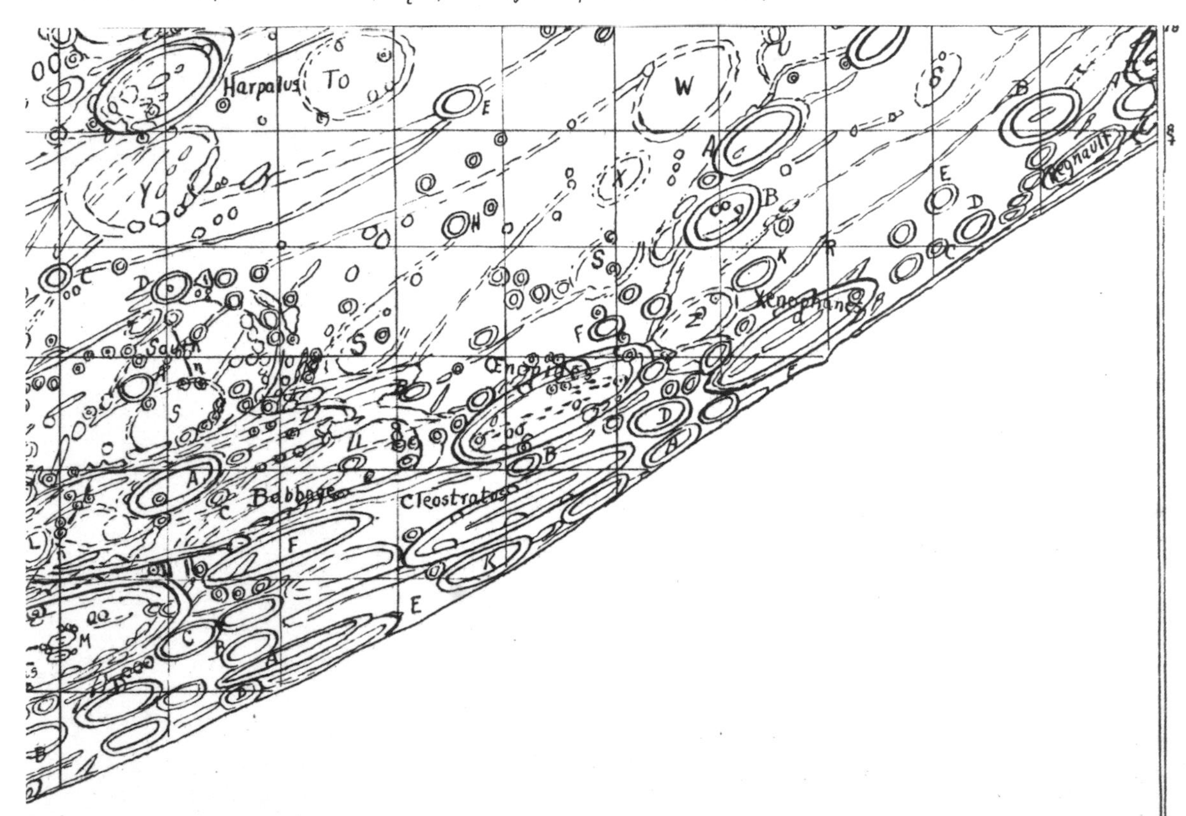

300 inch MAP OF THE MOON.

Section XVI.

Drawn by H. Percy Wilkins. F.R.A.S. 1946.

Director of the Lunar Section of the British Astronomical Association.

REPRODUCED BY W.F. STANLEY & Co. LTD. 13, RAILWAY APPROACH, LONDON BRIDGE S.E. 1.

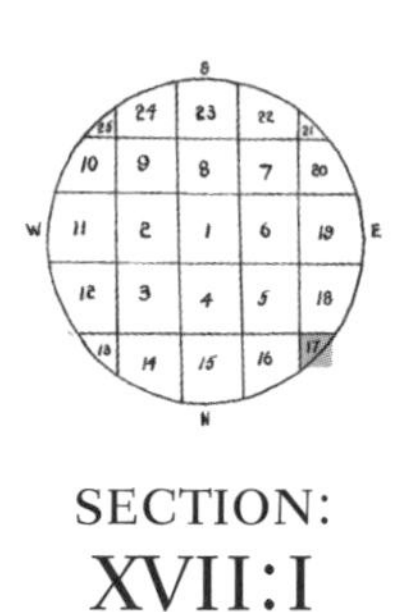

SECTION:
XVII:I

KEY:- ◉ Craterlet, ⊙ Craterpit, ⊕ Cratercone, o Hillock, Clefts, ═ Ridges, ······ Light Streaks, Dark Variable Spots.

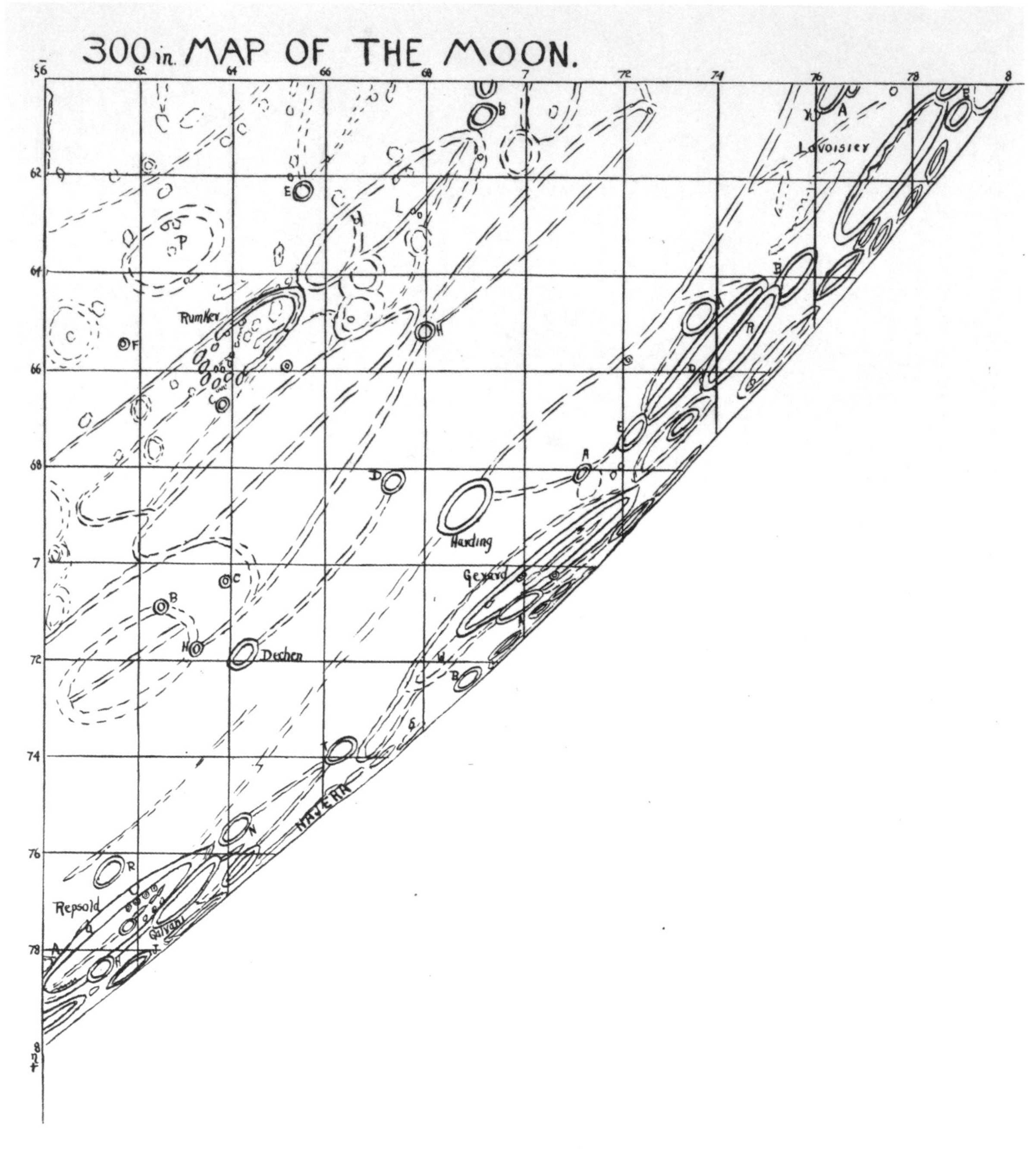

Mons Rūmker is an isolated lunar formation made up of a series of volcanic domes. Wilkins described it as a 'peculiar & fascinating object' in the notes which accompany one of his sketches of this formation, which can be found on page 113.

KEY:- ◎ Craterlet, ⊙ Craterpit, ⊕ Cratercone, o Hillock, Clefts, ═ Ridges, ······ Light Streaks, Dark Variable Spots.

SECTION. XVII.

8

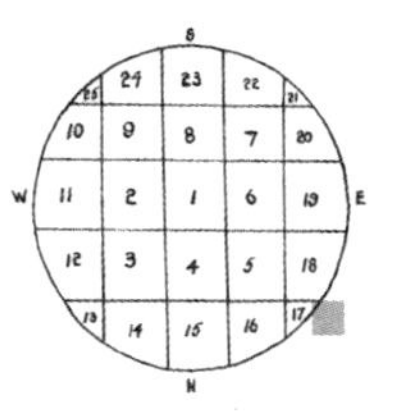

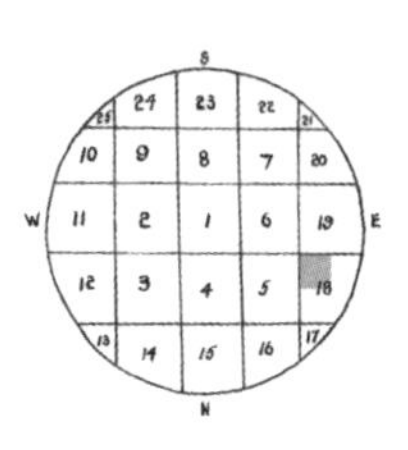

SECTION:
XVIII:I

KEY:- Craterlet, Craterpit, Cratercone, Hillock, Clefts, Ridges, Light Streaks, Dark Variable Spots.

300 in MAP OF THE MOON.

Oceanus Procellarum, the 'Ocean of Storms', is the largest of all the lunar 'seas' – so large, it is called an ocean. This ancient lava flow covers approximately 4 million km², about 16 times the area of the UK.

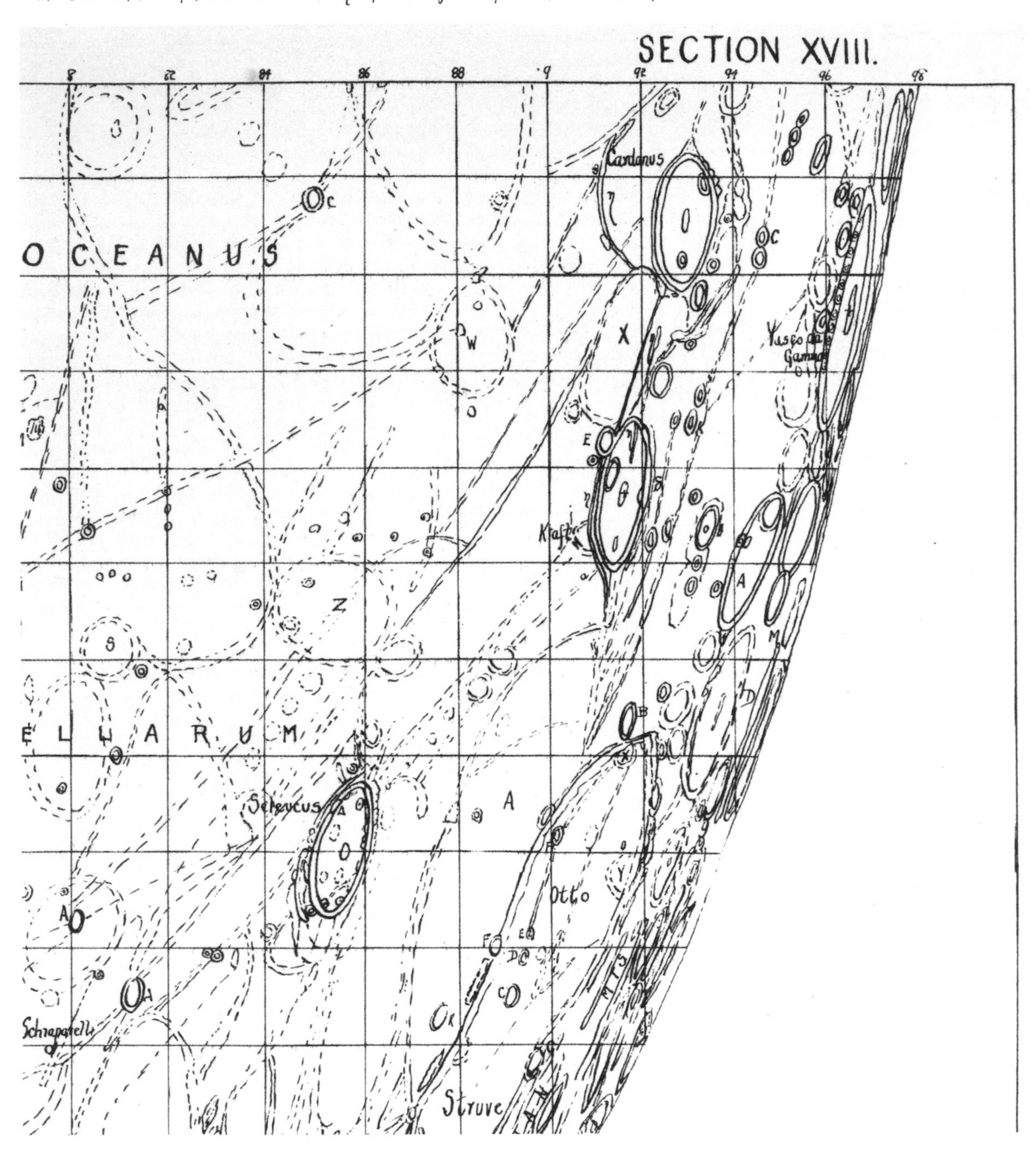
KEY:- Craterlet, Craterpit, Cratercone, Hillock, Clefts, Ridges, Light Streaks, Dark Variable Spots.
SECTION XVIII.
OCEANUS
Cardanus
Vasco da Gama
Krafft
Seleucus
Otto
Struve
Schiaparelli

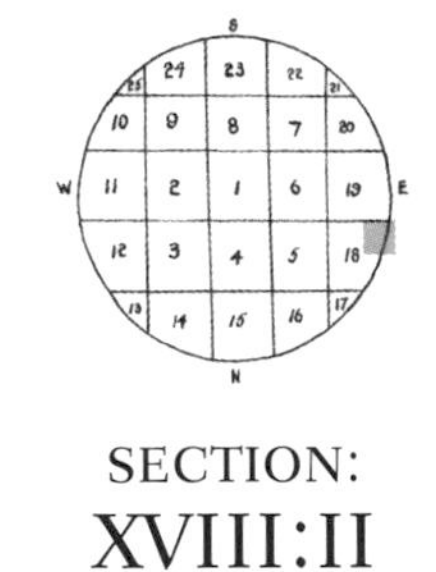

SECTION:
XVIII:II

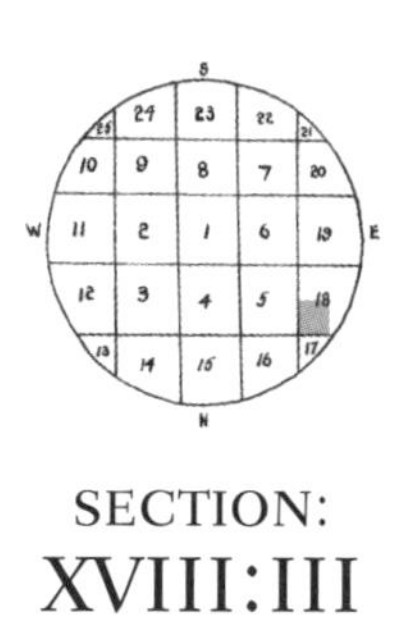
S
W
E
N

SECTION:
XVIII:III

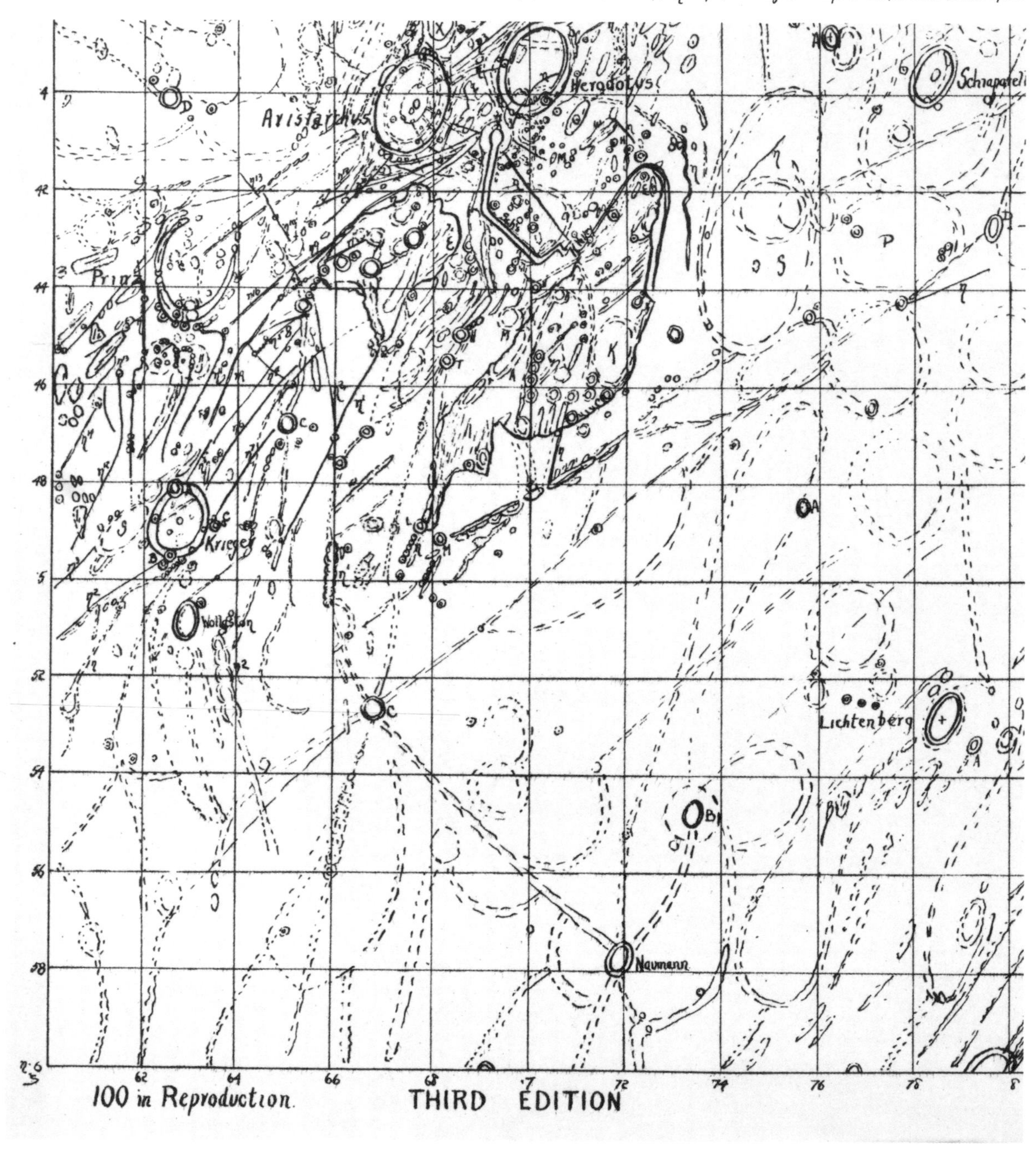
KEY:- Craterlet, Craterpit, Cratercone, Hillock, Clefts, Ridges, Light Streaks, Dark Variable Spots.
Aristarchus
Herodotus
Schiapateli
Prinz
Krieger
Wollaston
Lichtenberg
Naumann
100 in Reproduction.
THIRD EDITION

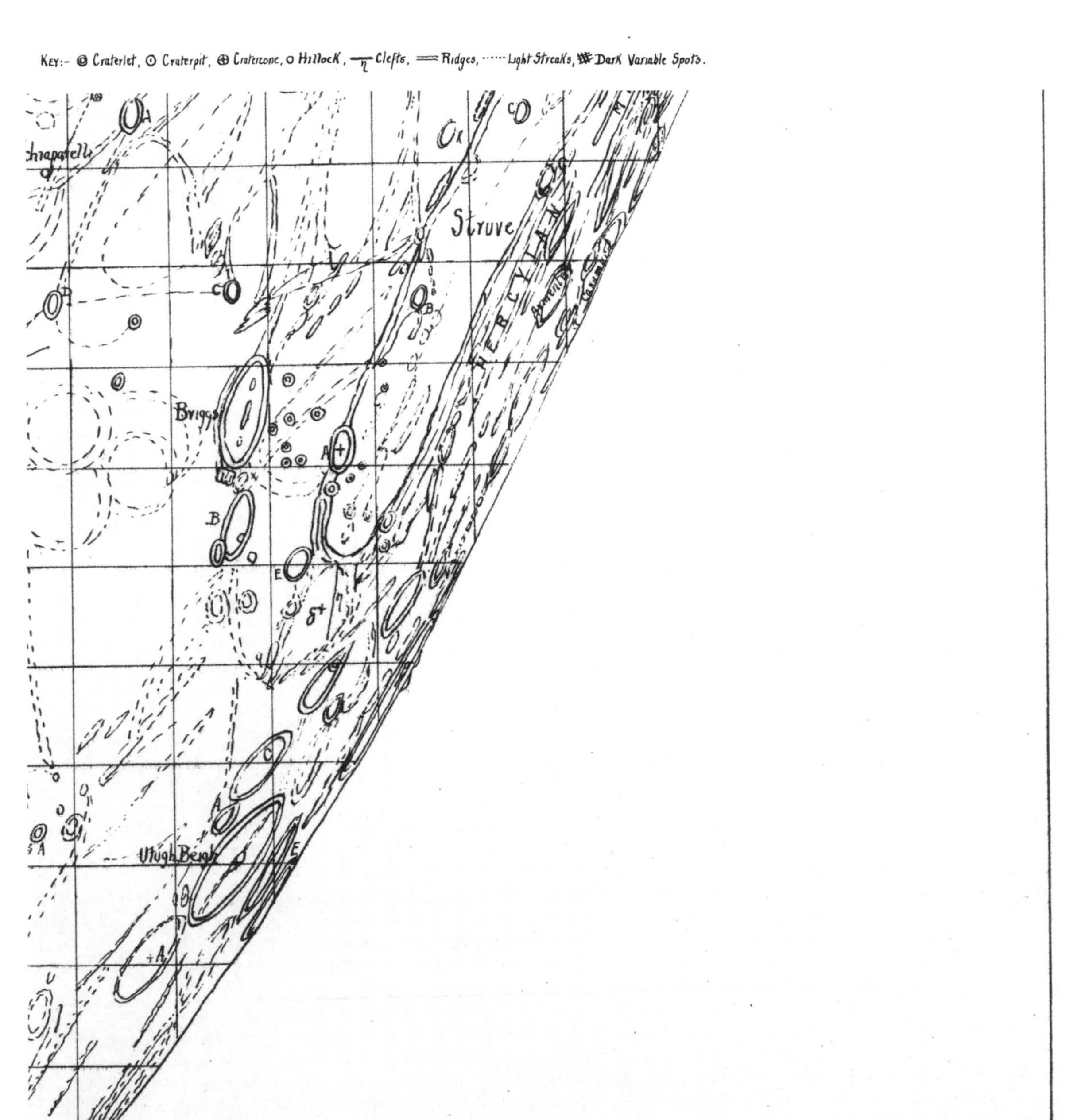

Drawn by H. Percy Wilkins. F.R.A.S. 1951.

PRINTED BY W. F. STANLEY & Co LTD 13 RAILWAY APPROACH, LONDON BRIDGE S.E.1.

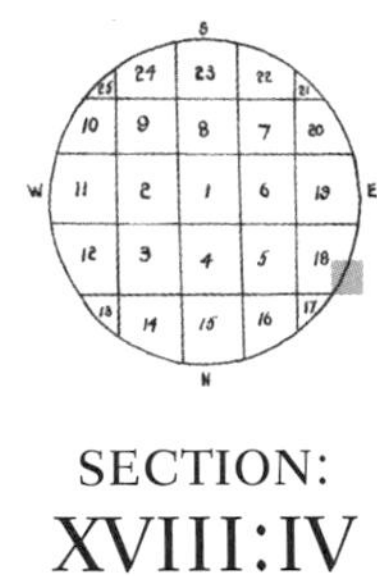

SECTION:
XVIII:IV

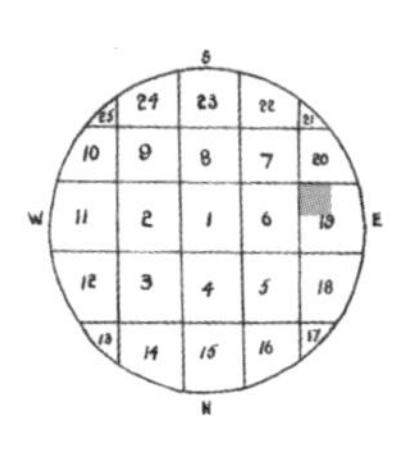

SECTION:
XIX:I

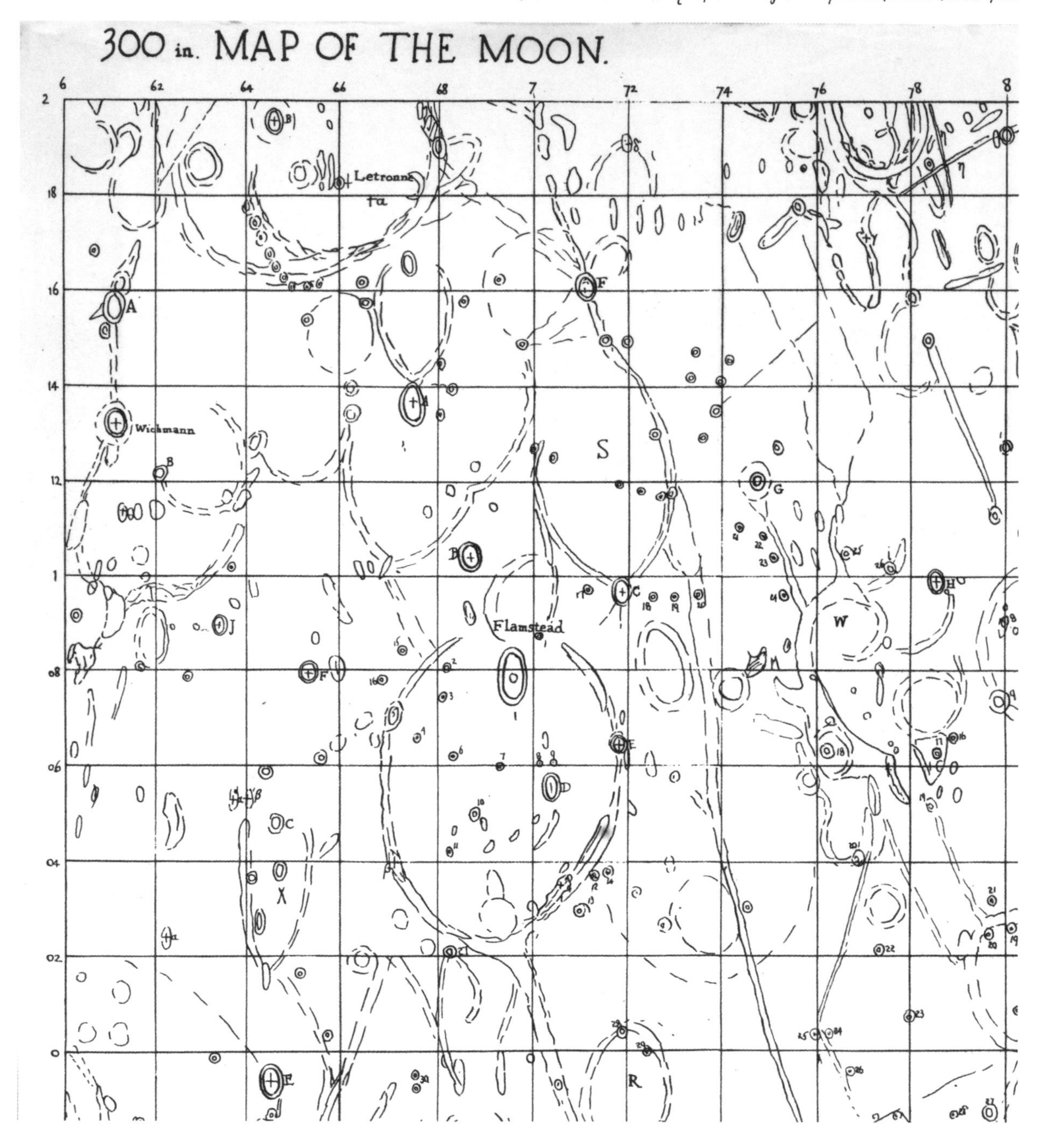

At the centre of this sheet is a crater named after **John Flamsteed** (1646–1719), who was the first Astronomer Royal to serve at the Royal Observatory, Greenwich. He was appointed by King Charles II to make a map of the heavens reliable enough to be useful for navigators.

KEY:– ◎ Craterlet, ⊙ Craterpit, ⊕ Cratercone, ○ Hillock, —η— Clefts, ═ Ridges, ······ Light Streaks, ## Dark Variable Spots.

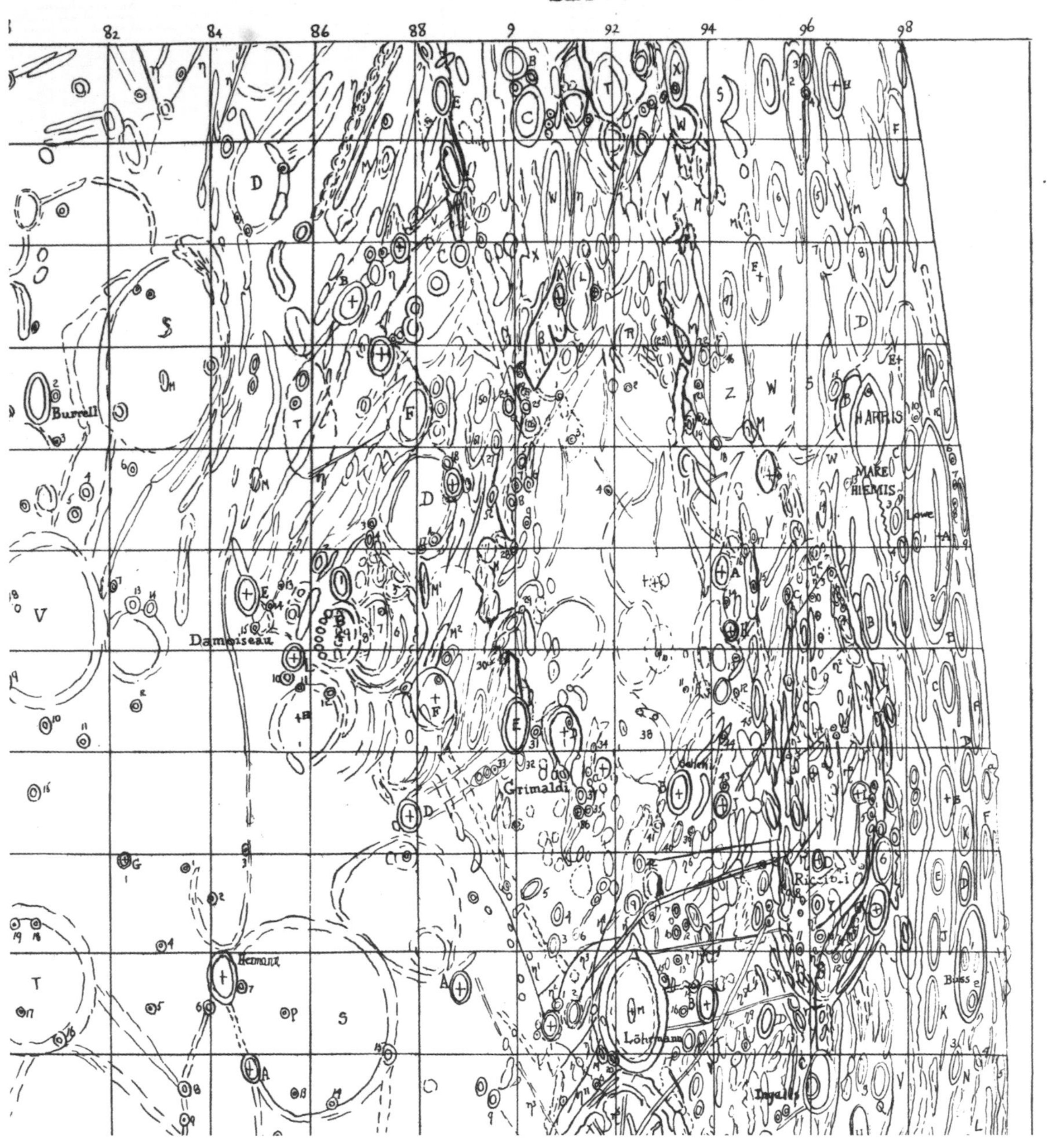

SECTION:
XIX:II

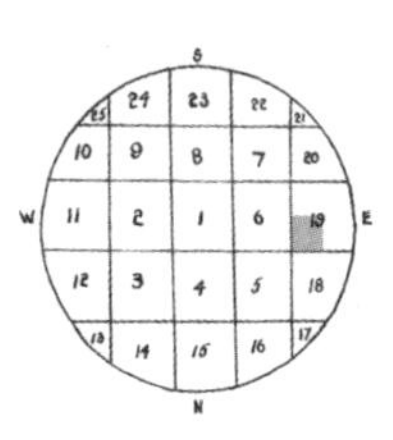

SECTION:
XIX:III

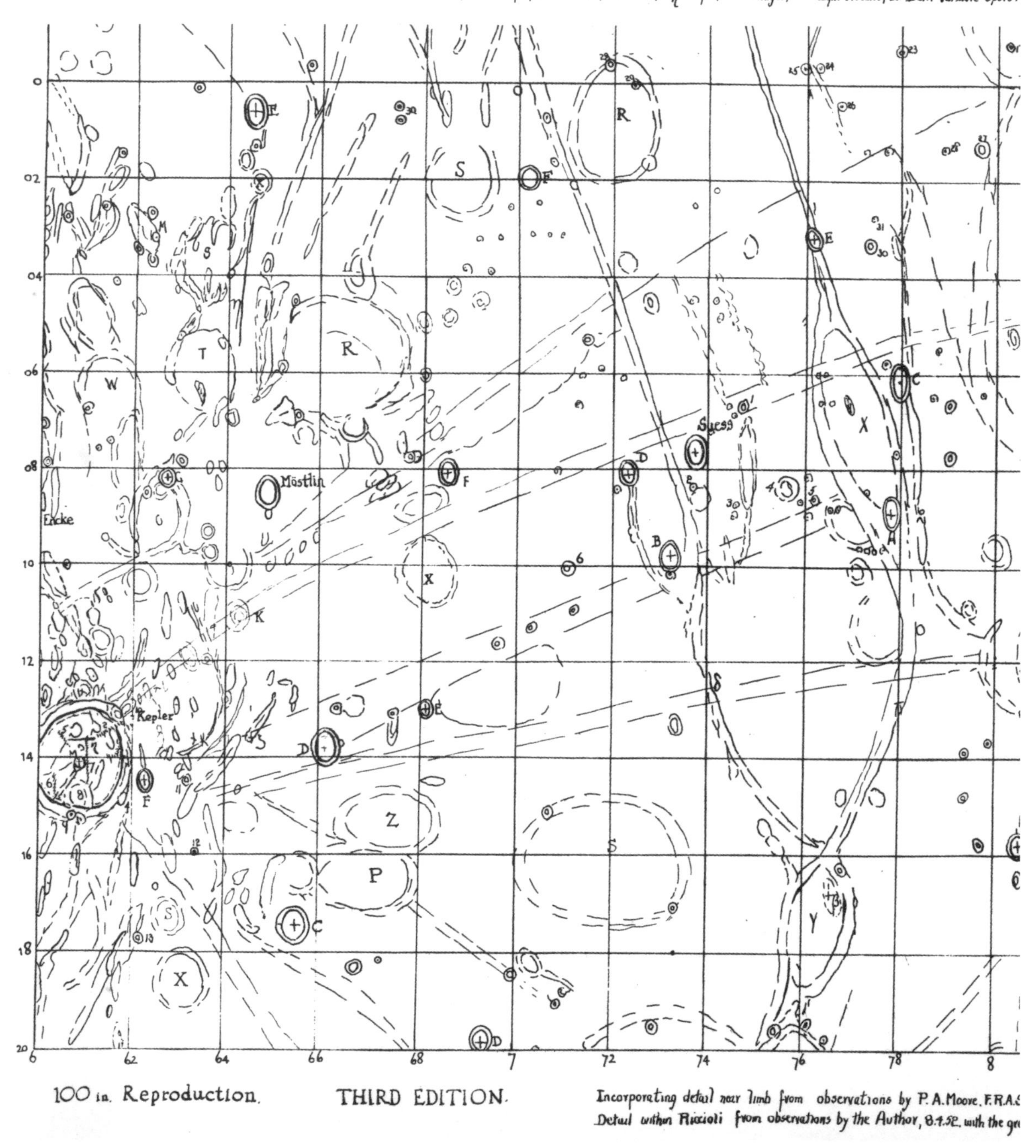

Bright, reflective rays surround many craters on the Moon. On this sheet, they be seen clearly spreading out from the crater **Kepler**. Rays form when an asteroid impact scatters streaks of debris across the Moon's surface. Extensive systems of rays like this are features of newer craters; the rays of older craters tend to be destroyed by later asteroid impacts.

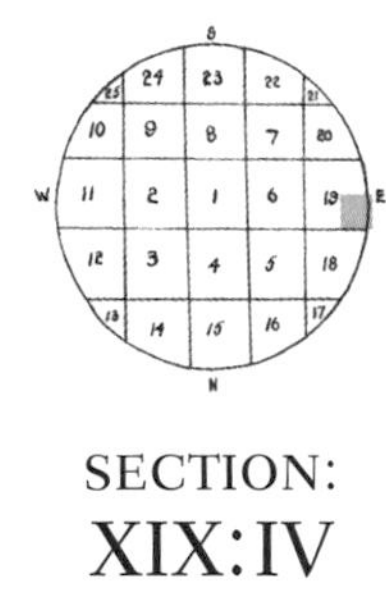

SECTION:
XIX:IV

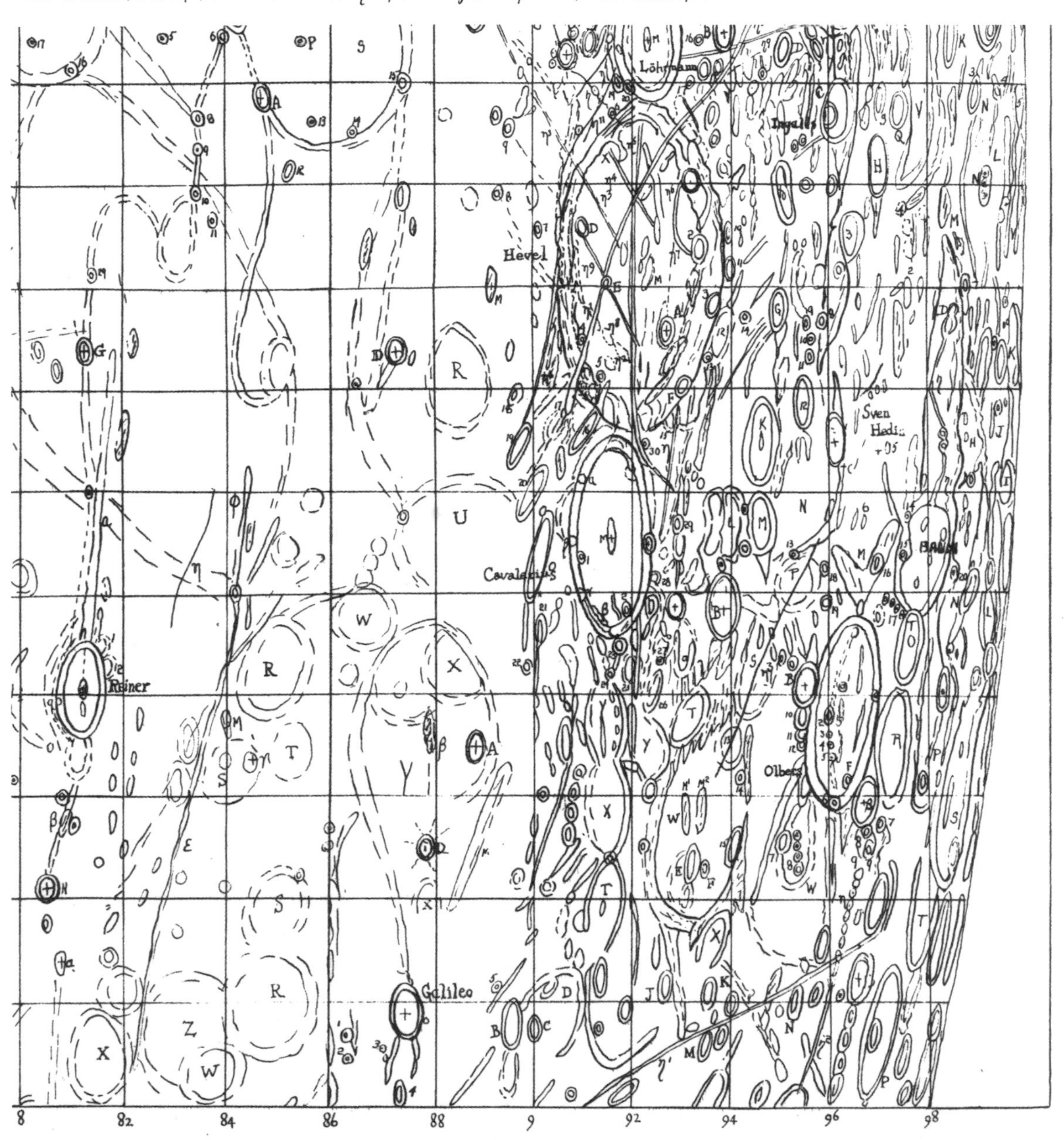

Drawn by H. Percy Wilkins. F.R.A.S. 1951.

PRINTED BY W. F. STANLEY & Co LTD 13. RAILWAY APPROACH, LONDON BRIDGE S.E.1.

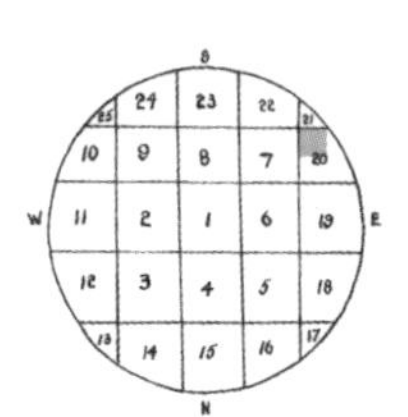

SECTION:
XX:I

KEY:- ◎ Craterlet, ⊙ Craterpit, ⊕ Cratercone, ○ Hillock, —η— Clefts, ═ Ridges, ······ Light Streaks, ▦ Dark Variable Spots.

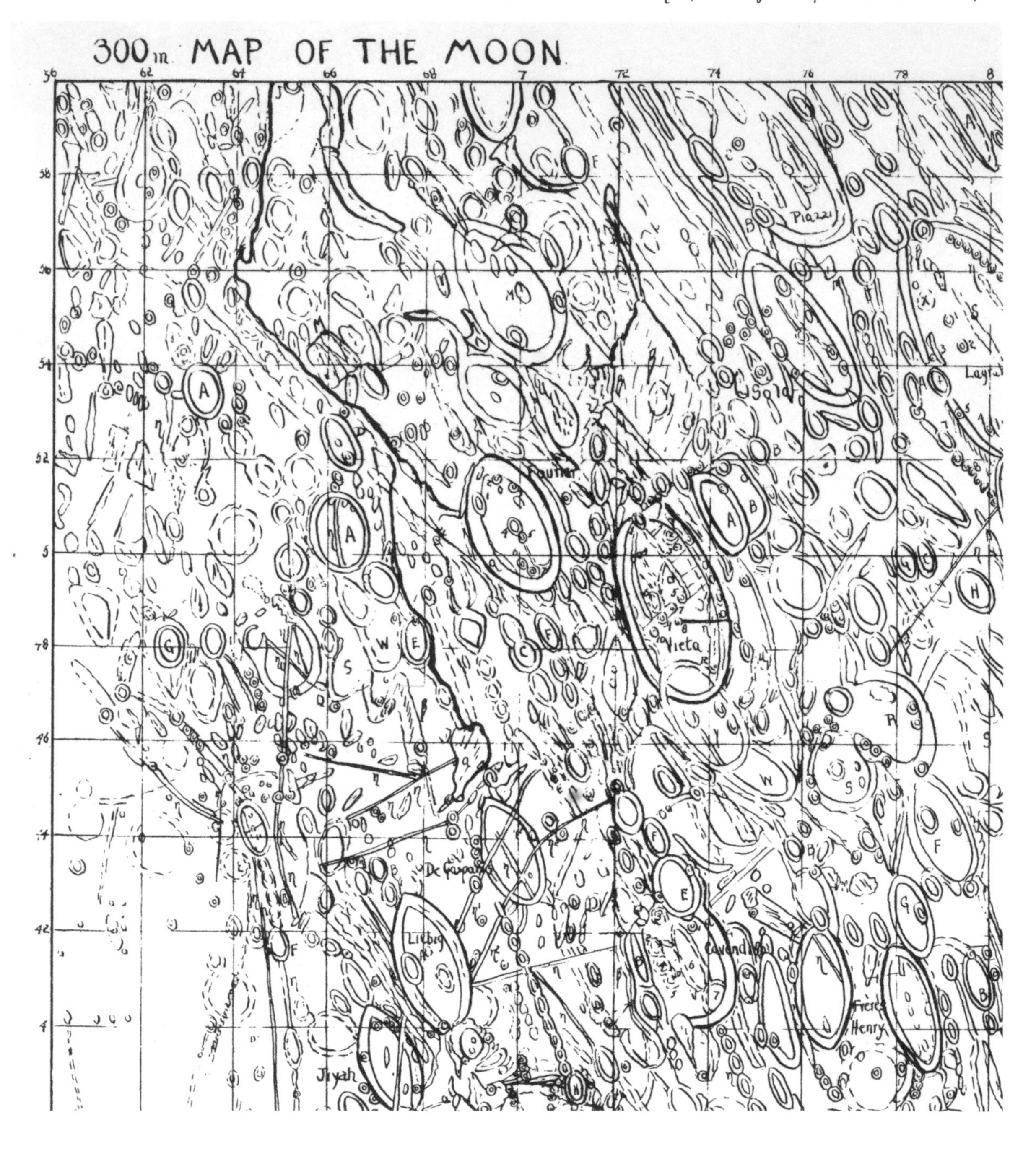

Key:- ◎ Craterlet, ⊙ Craterpit, ⊕ Cratercone, ○ Hillock, —η— Clefts, ══ Ridges, ······ Light Streaks, # Dark Variable Spots.

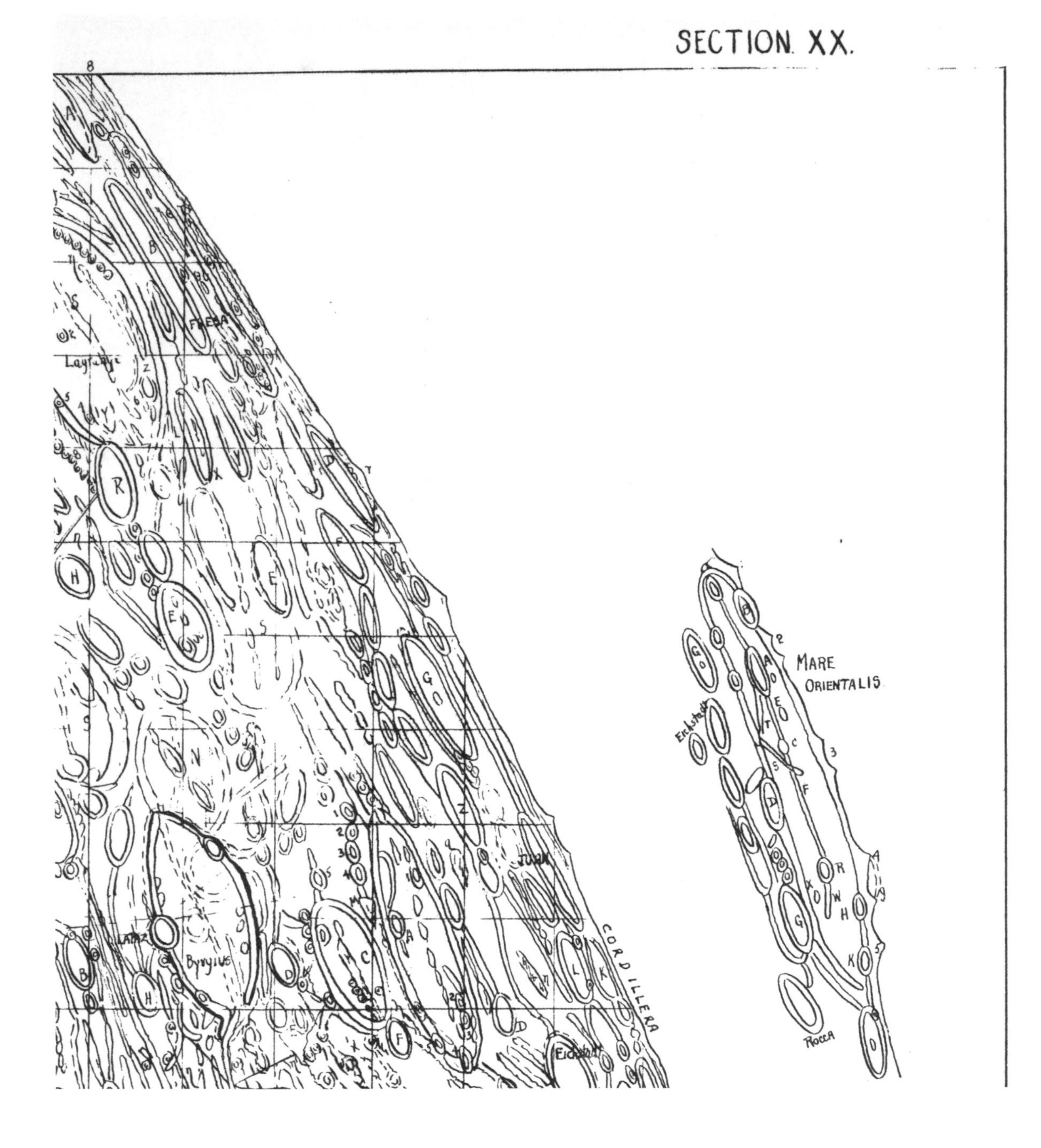

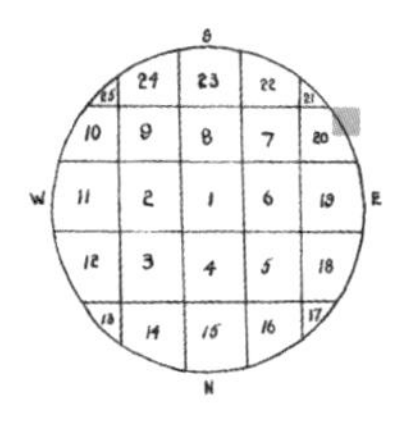

SECTION:
XX:II

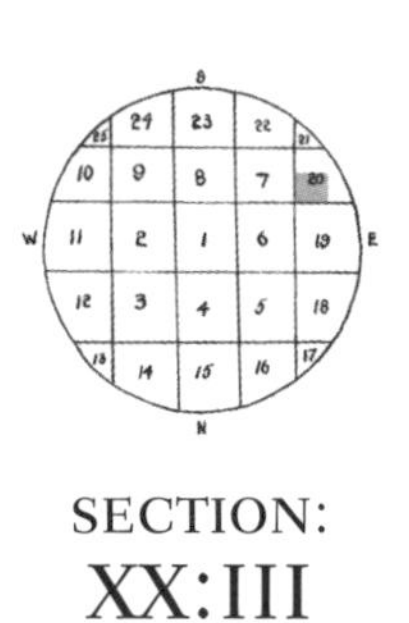

SECTION:
XX:III

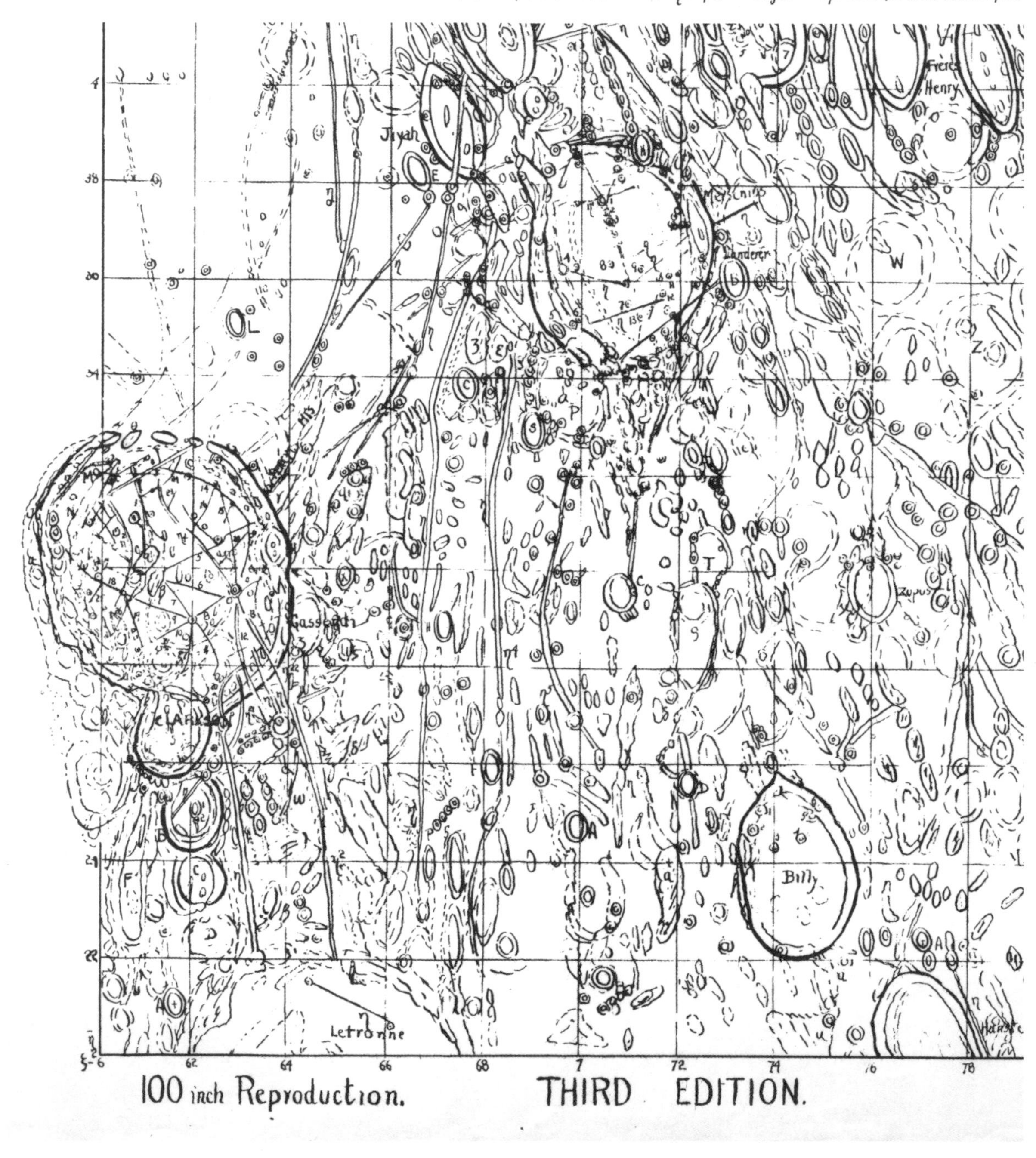

Many larger craters on the Moon have peaks at their centre. In the crater **Gassendi**, this is marked by the circles in the centre. Scientists now think these are caused by rebound after asteroid impact, as the underlying crust bounces back and pushes up a mass of rock.

KEY:- ◎ Craterlet, ⊙ Craterpit, ⊕ Cratercone, o Hillock, —η— Clefts, ═══ Ridges, ······ Light Streaks, ▦ Dark Variable Spots.

SECTION:
XX:IV

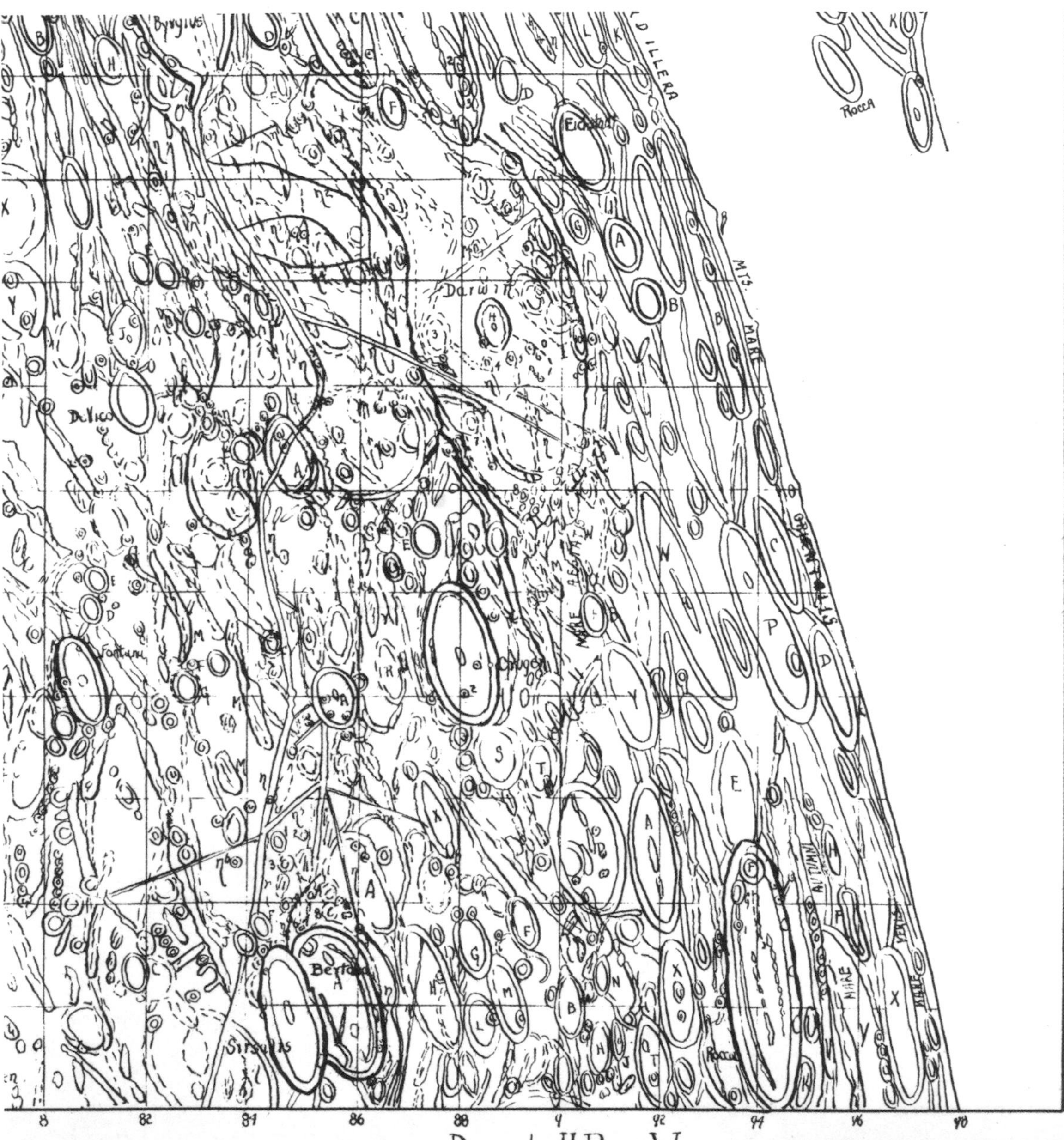

Drawn by H. PERCY WILKINS F.R.A.S. 1951.

PRINTED BY W. F. STANLEY & Co. LTD 13. RAILWAY APPROACH, LONDON BRIDGE. S.E.1.

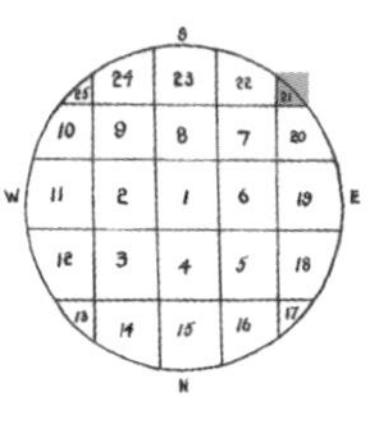

SECTION:
XXI:I

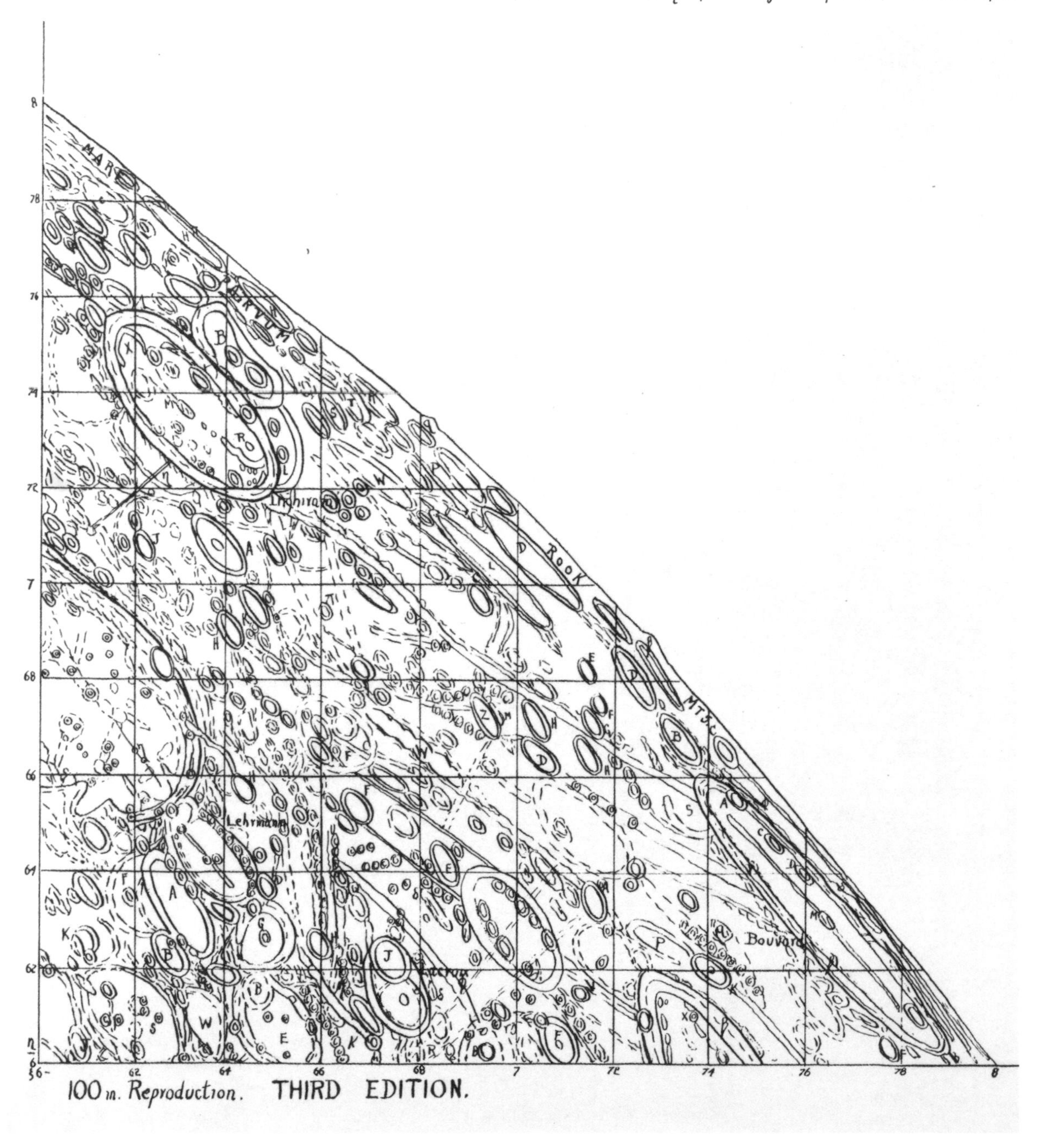

The impact which formed the **Mare Orientale** (which Wilkins called Mare Orientalis) was so big it created concentric rings around the impact site. The Montes Rook (to which Wilkins gives their English name, the Rook Mountains) are one of these rings. Part of the range is on the far side of the Moon and is not visible form Earth.

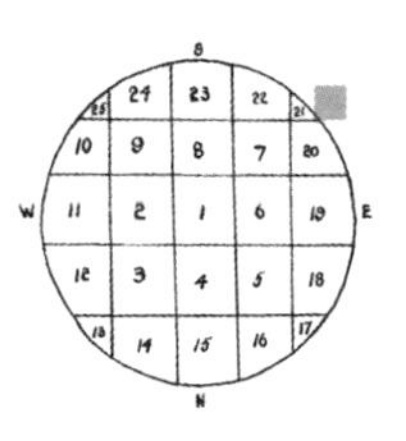

8

Drawn by H. Percy Wilkins. F.R.A.S.- 1951

PRINTED BY W. F. STANLEY & CO., LTD. 13, RAILWAY APPROACH, LONDON BRIDGE, S.E.1.

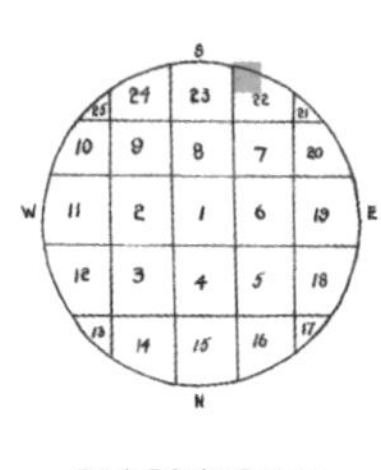

SECTION:
XXII:I

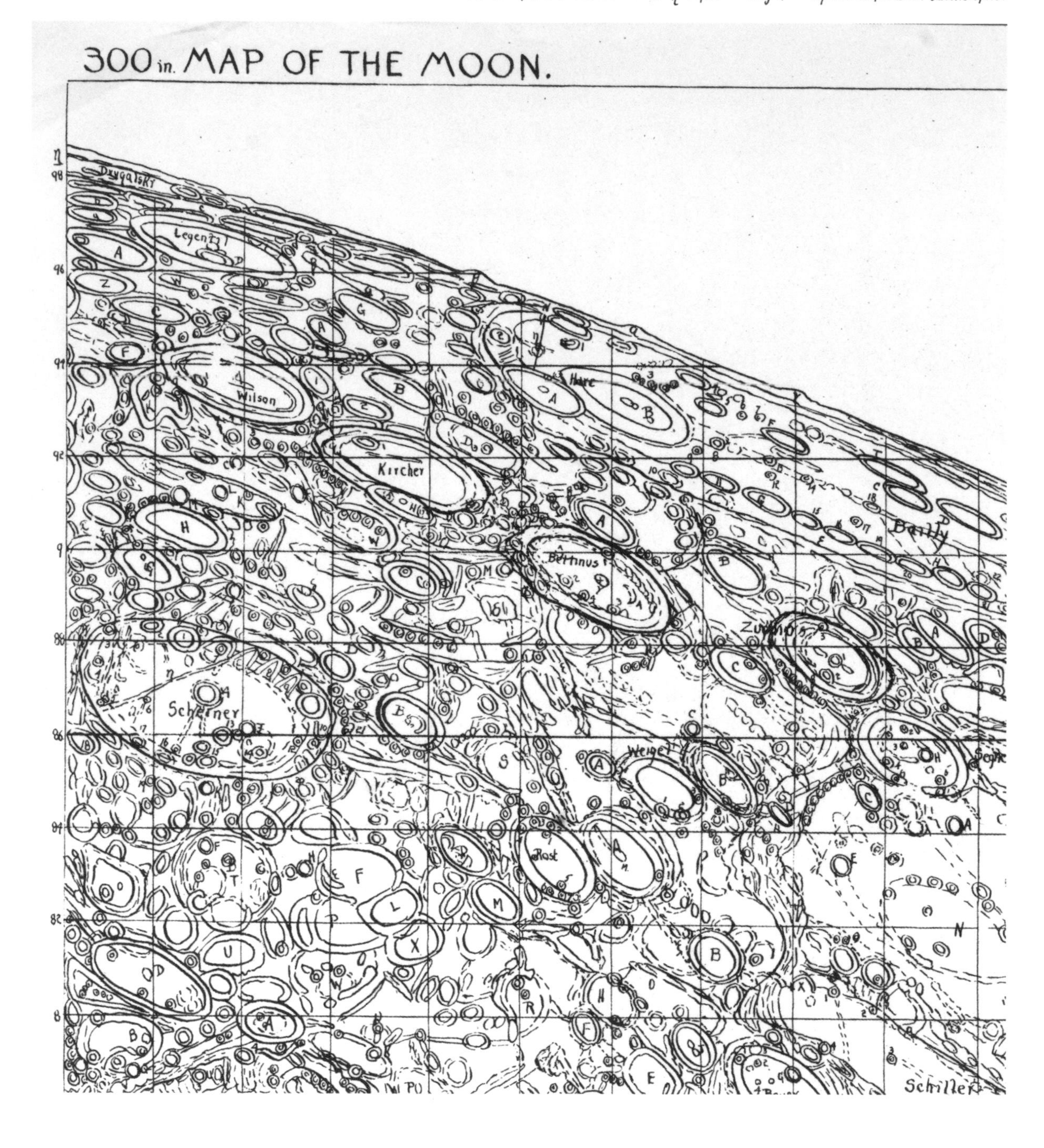
KEY:- Craterlet, Craterpit, Cratercone, Hillock, Clefts, Ridges, Light Streaks, Dark Variable Spots.
300 in. MAP OF THE MOON.
Legentzl
Wilson
Kircher
Hare
Bettinus
Bailly
Zucchius
Scheiner
Weigel
Rost
Schiller

KEY:- ◎ Craterlet, ⊙ Craterpit, ⊕ Cratercone, o Hillock, Clefts, ═ Ridges, ····· Light Streaks, Dark Variable Spots.

SECTION. XXII.

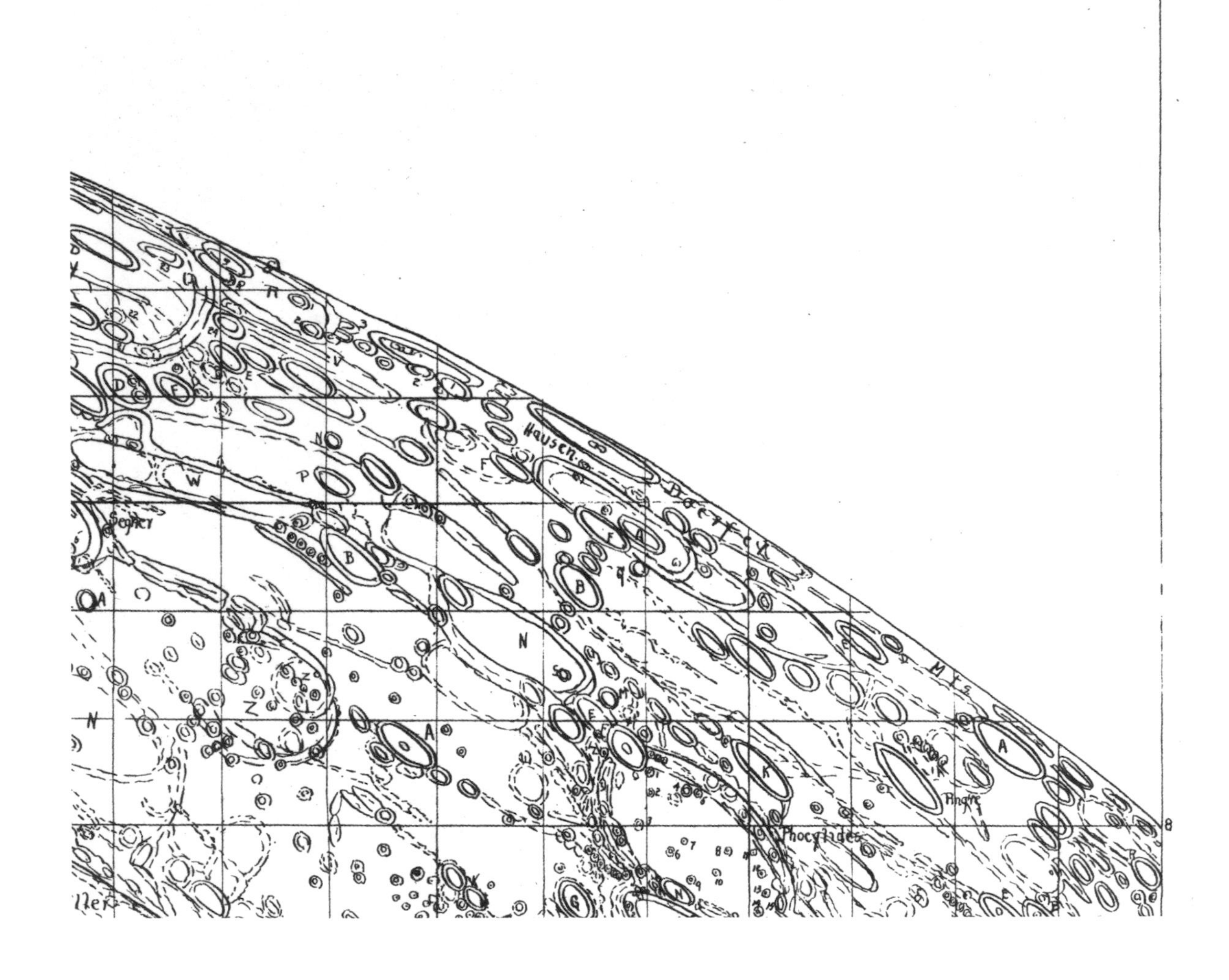

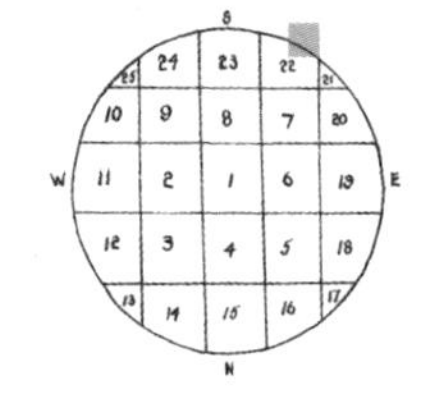

SECTION:
XXII:II

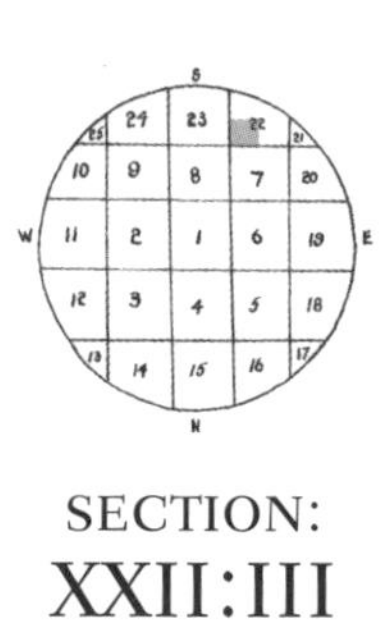

SECTION:
XXII:III

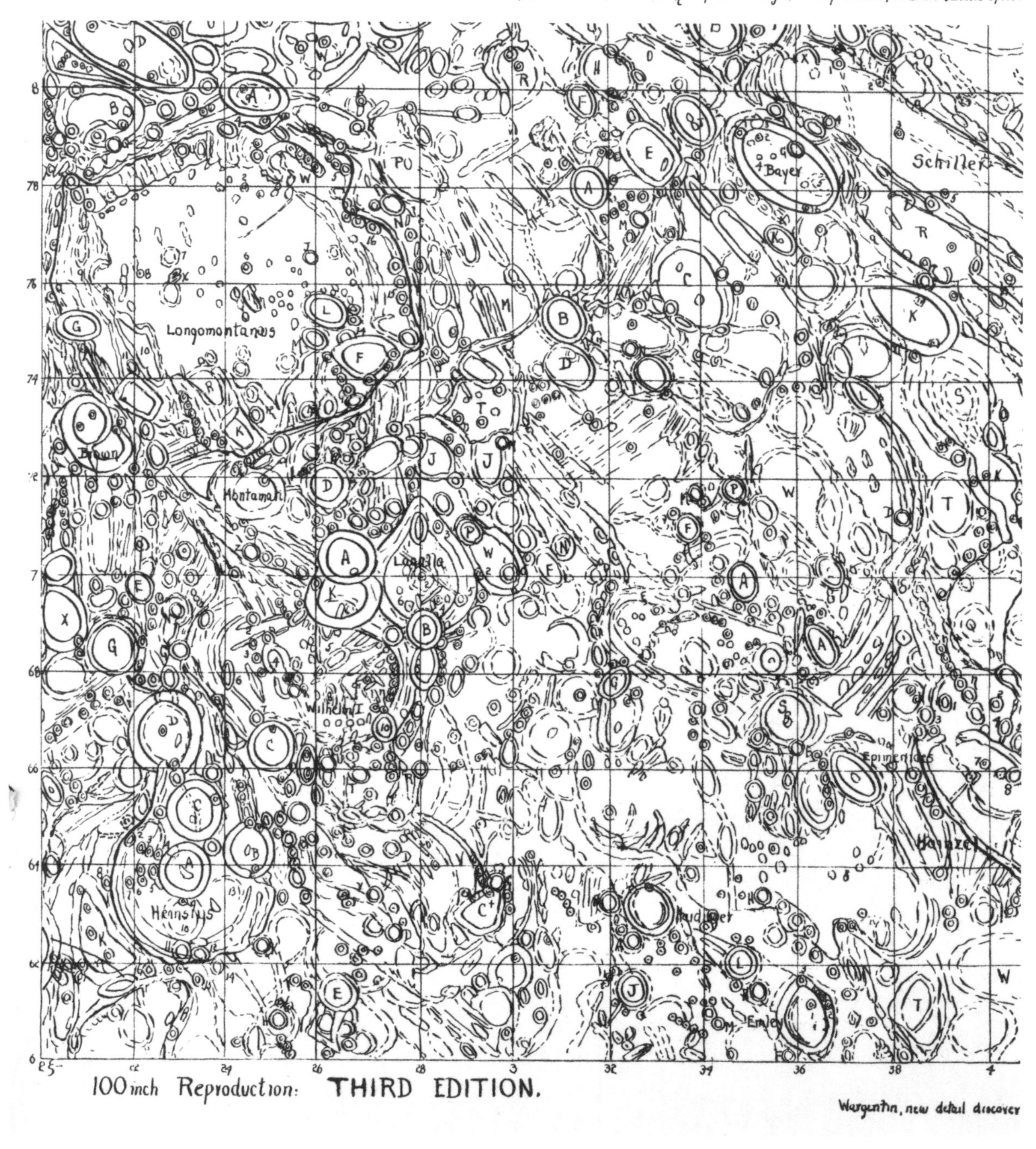

At 179 km long and 71 km wide, **Schiller** is an unusually elongated crater. This may be the result of multiple impacts, or an impact which 'grazed' the surface of the Moon.

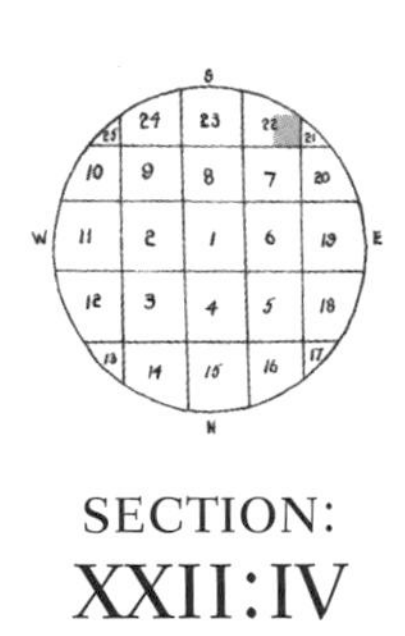
24
23
22
21
10
9
8
7
20
W
11
2
1
6
19
E
12
3
4
5
18
13
14
15
16
17
N

SECTION:
XXII:IV

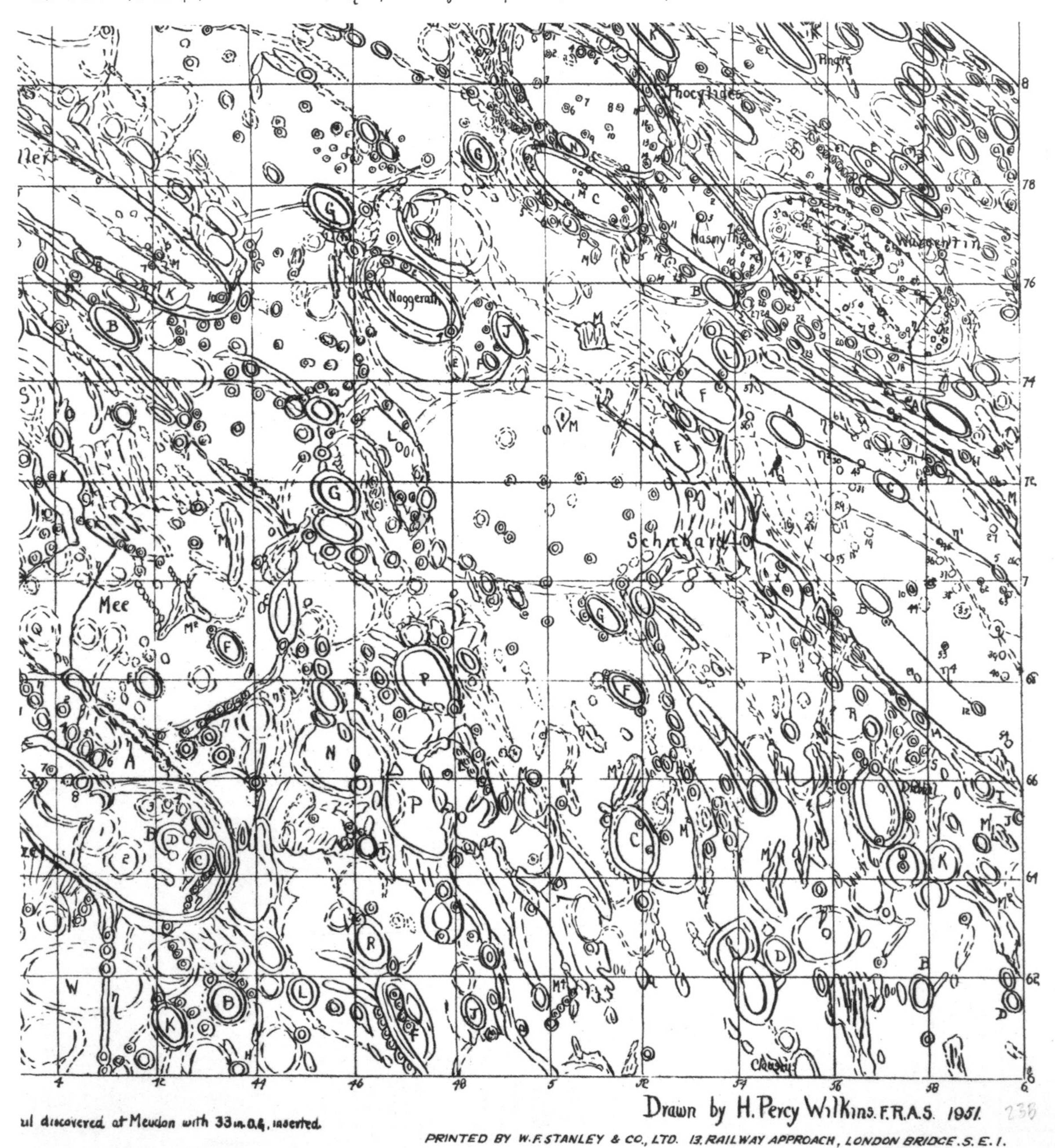
KEY:- Craterlet, Craterpit, Cratercone, Hillock, Clefts, Ridges, Light Streaks, Dark Variable Spots.
Phocylides
Noggerath
Nasmyth
Wargentin
Schickard
Mee
Drawn by H. Percy Wilkins. F.R.A.S. 1951.
...ul discovered at Meudon with 33 in. O.G. inserted.
PRINTED BY W.F. STANLEY & CO., LTD. 13, RAILWAY APPROACH, LONDON BRIDGE. S.E.1.

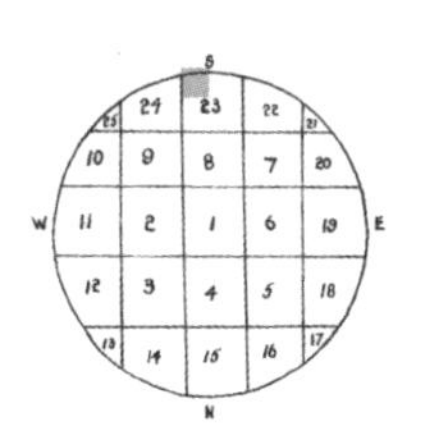

SECTION:
XXIII:I

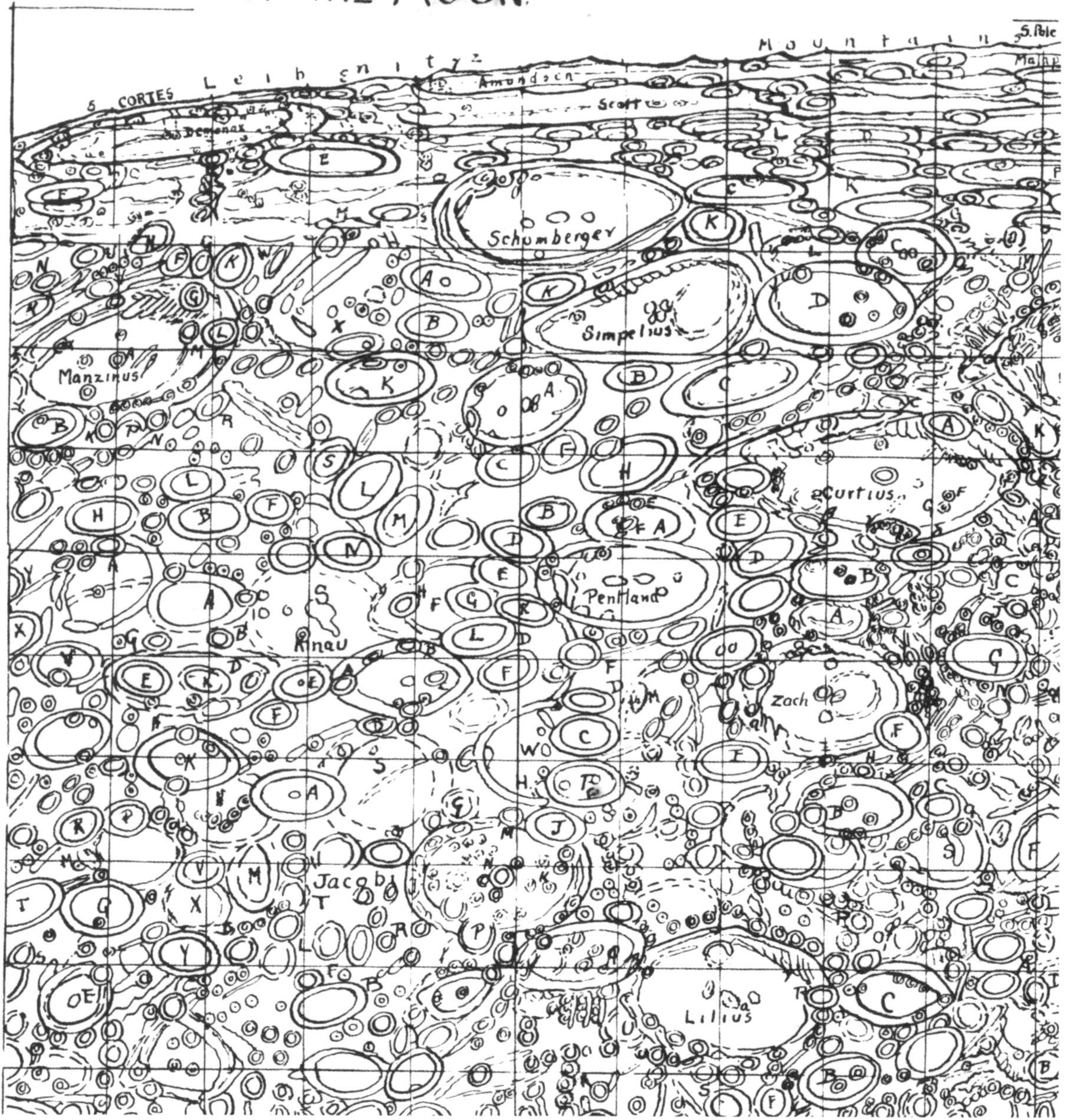

Robert Falcon Scott and **Roald Amundsen** set out in 1911 to race to the South Pole. Famously, Amundsen got there first. Tragically, Scott and his companions never made it back. Two craters near the lunar South Pole were named after these men. Amundsen is slightly closer to the Pole.

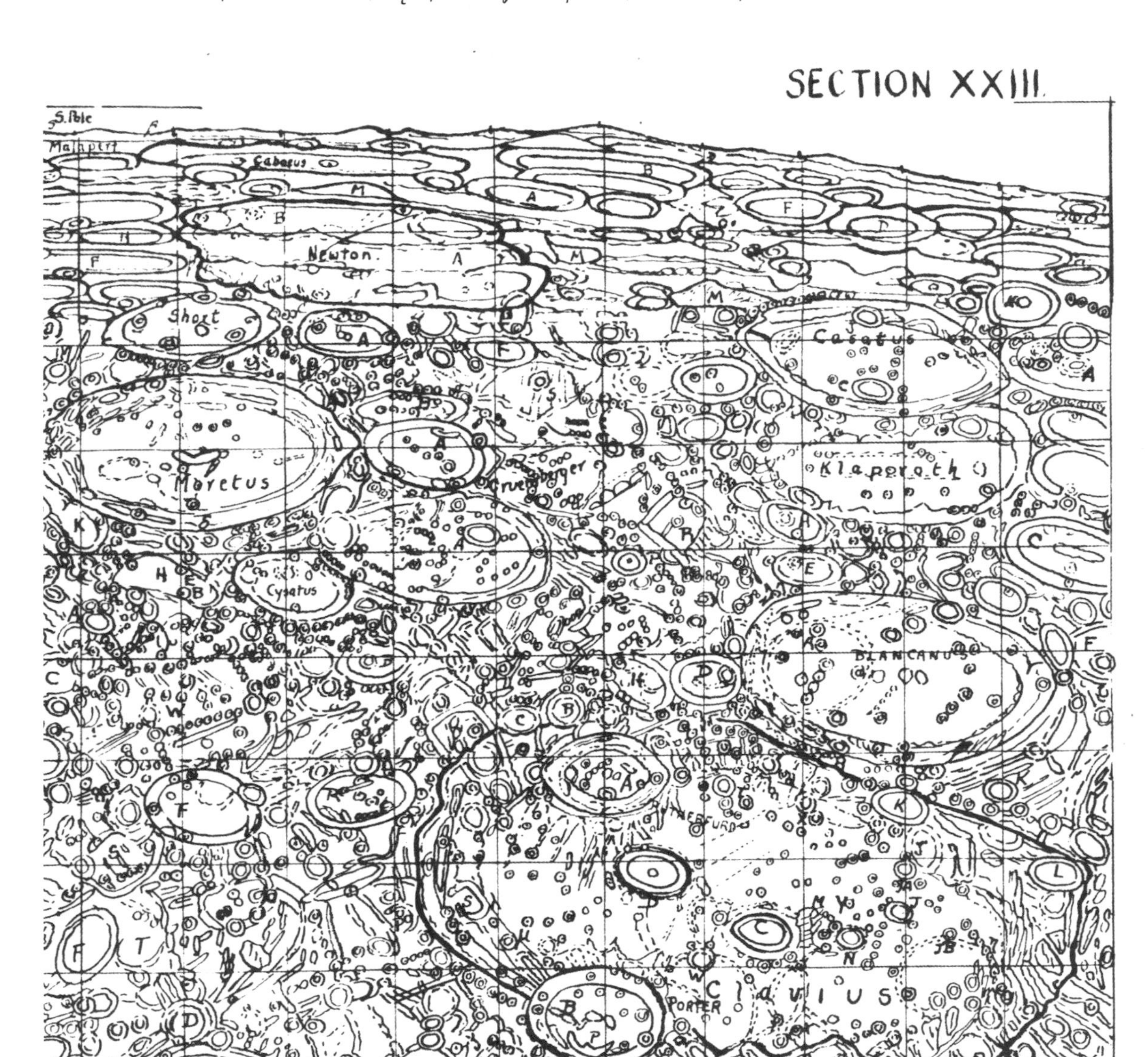

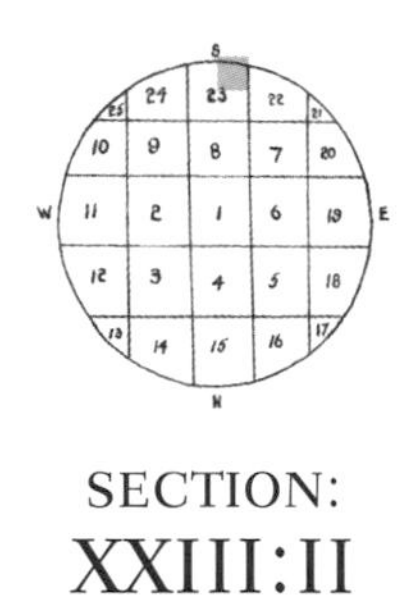

SECTION:
XXIII:II

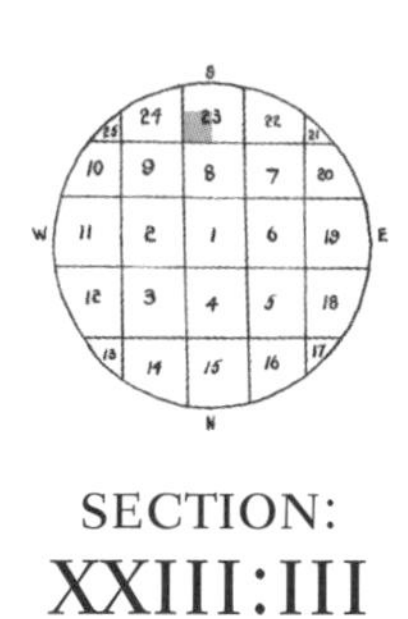

SECTION:
XXIII:III

KEY:- ◎ Craterlet, ⊙ Craterpit, ⊕ Cratercone, o Hillock, Clefts, ═ Ridges, ... Light Streaks, Dark Variable Spots.

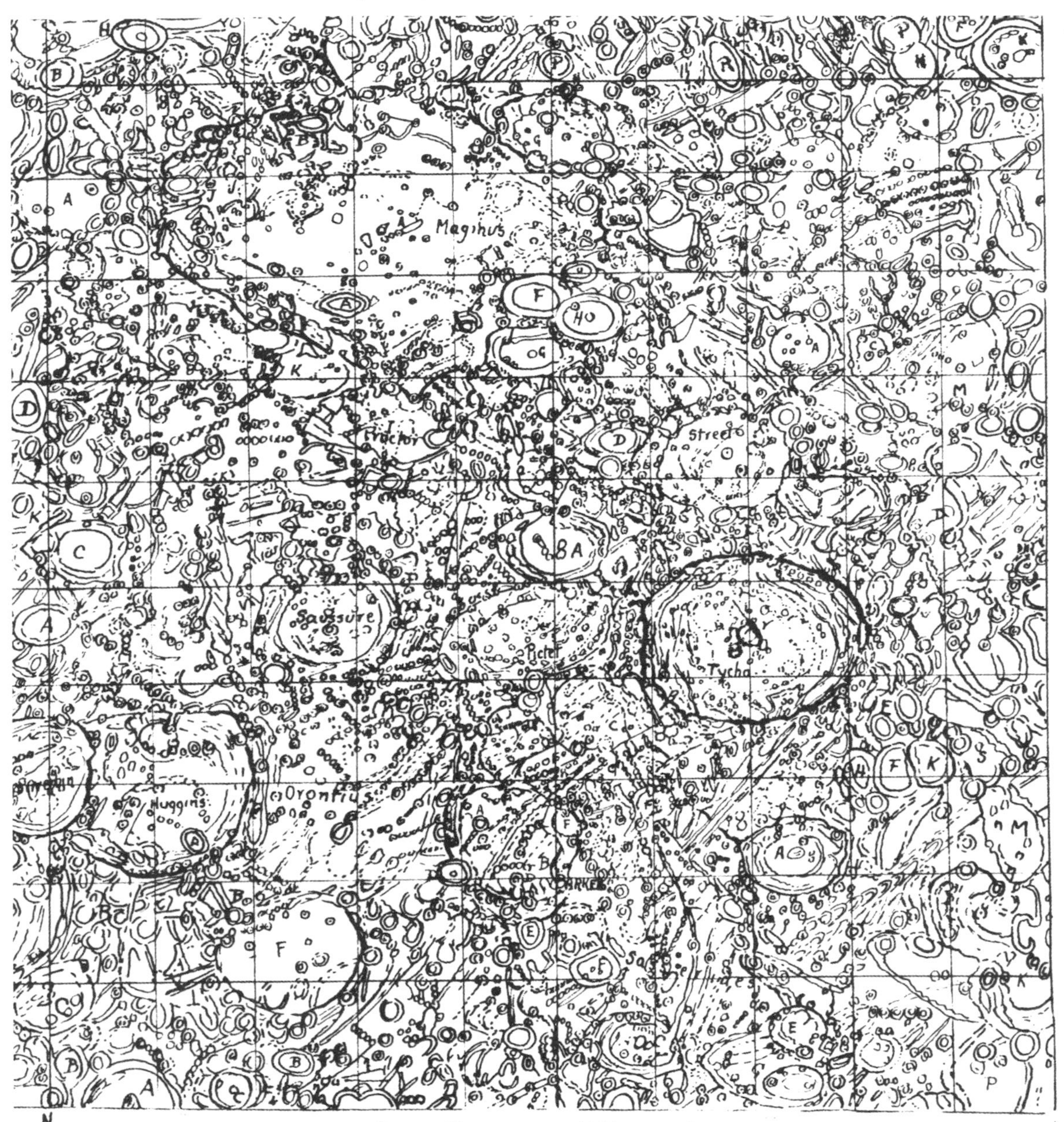

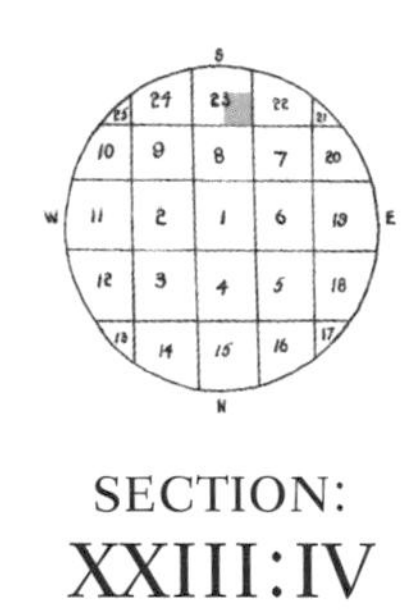

SECTION:
XXIII:IV

Maginus I = Proctor. Sasserides B = Barker: Clavius B = Porter. DRAWN BY H. PERCY WILKINS. 1937.

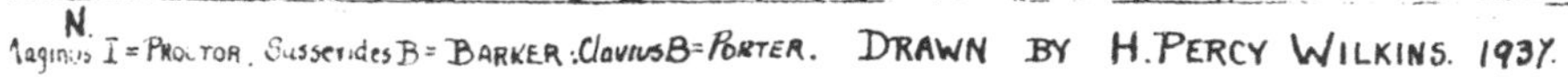
PRINTED BY W. F. STANLEY & Co. LTD 13, RAILWAY APPROACH, LONDON BRIDGE. S.E.1.

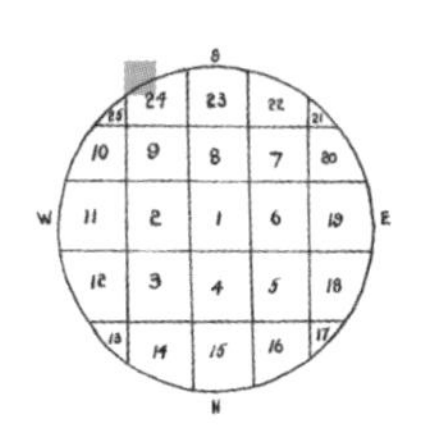

SECTION:
XXIV:I

KEY:- ◎ Craterlet, ⊙ Craterpit, ⊕ Cratercone, ○ Hillock, Clefts, Ridges, ... Light Streaks, Dark Variable Spots.

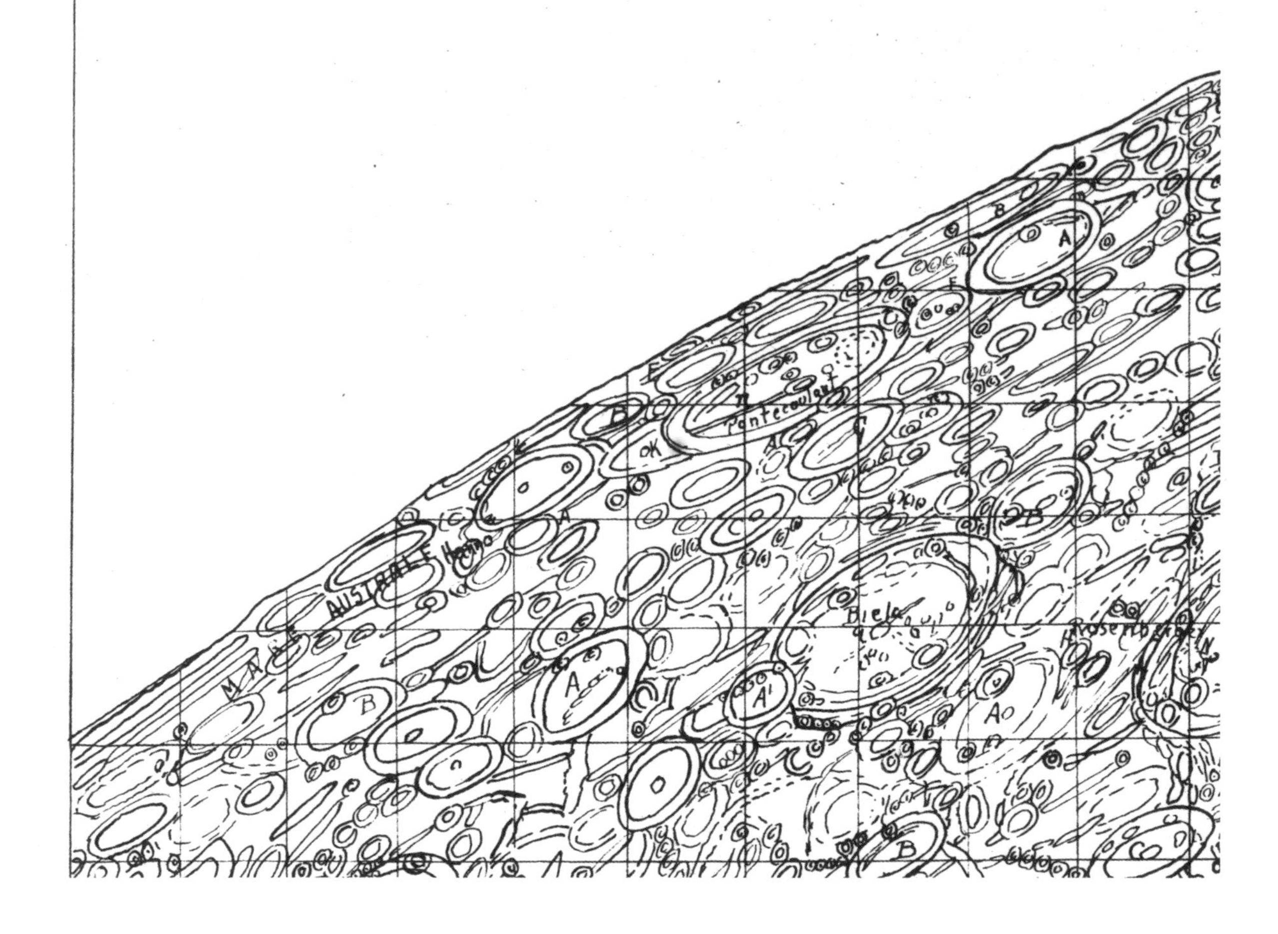

Key:- ◎ Craterlet, ⊙ Craterpit, ⊕ Cratercone, ○ Hillock, Clefts, = Ridges, ······ Light Streaks, Dark Variable Spots.

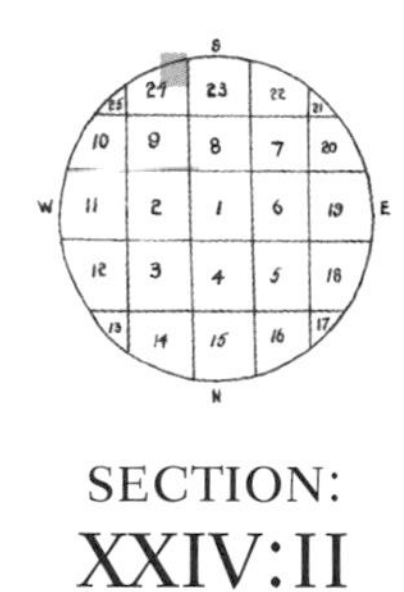

SECTION:
XXIV:II

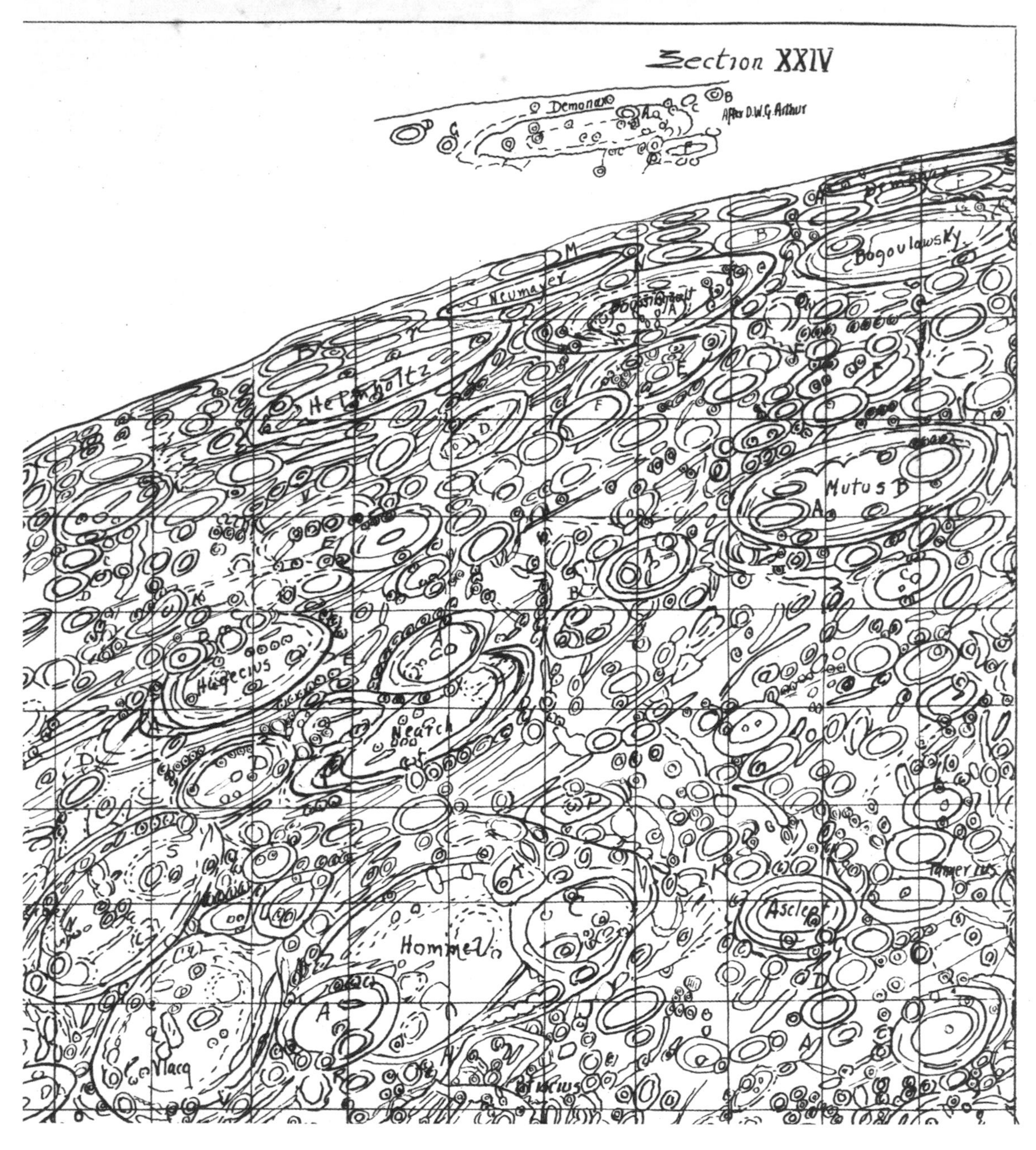

Wilkins's map incorporated cartographic details from other astronomers, for instance in the detailed view of the crater **Demonax**, 'after D.W.G Arthur.' David Arthur, a Welsh astronomer who was part of the Lunar Section of the British Astronomical Association which Wilkins directed, would go on to work at the Lunar and Planetary Laboratory at the University of Arizona.

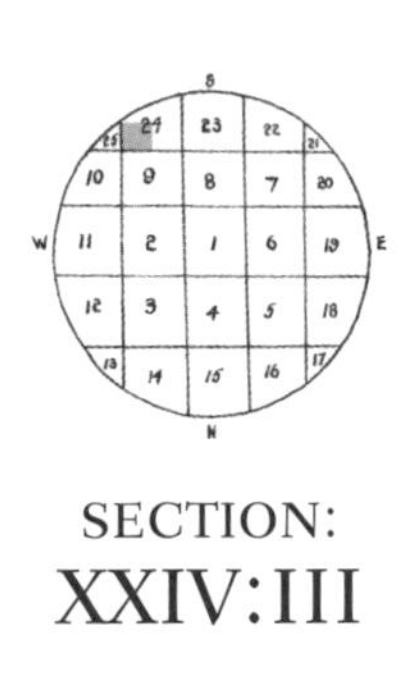

SECTION:
XXIV:III

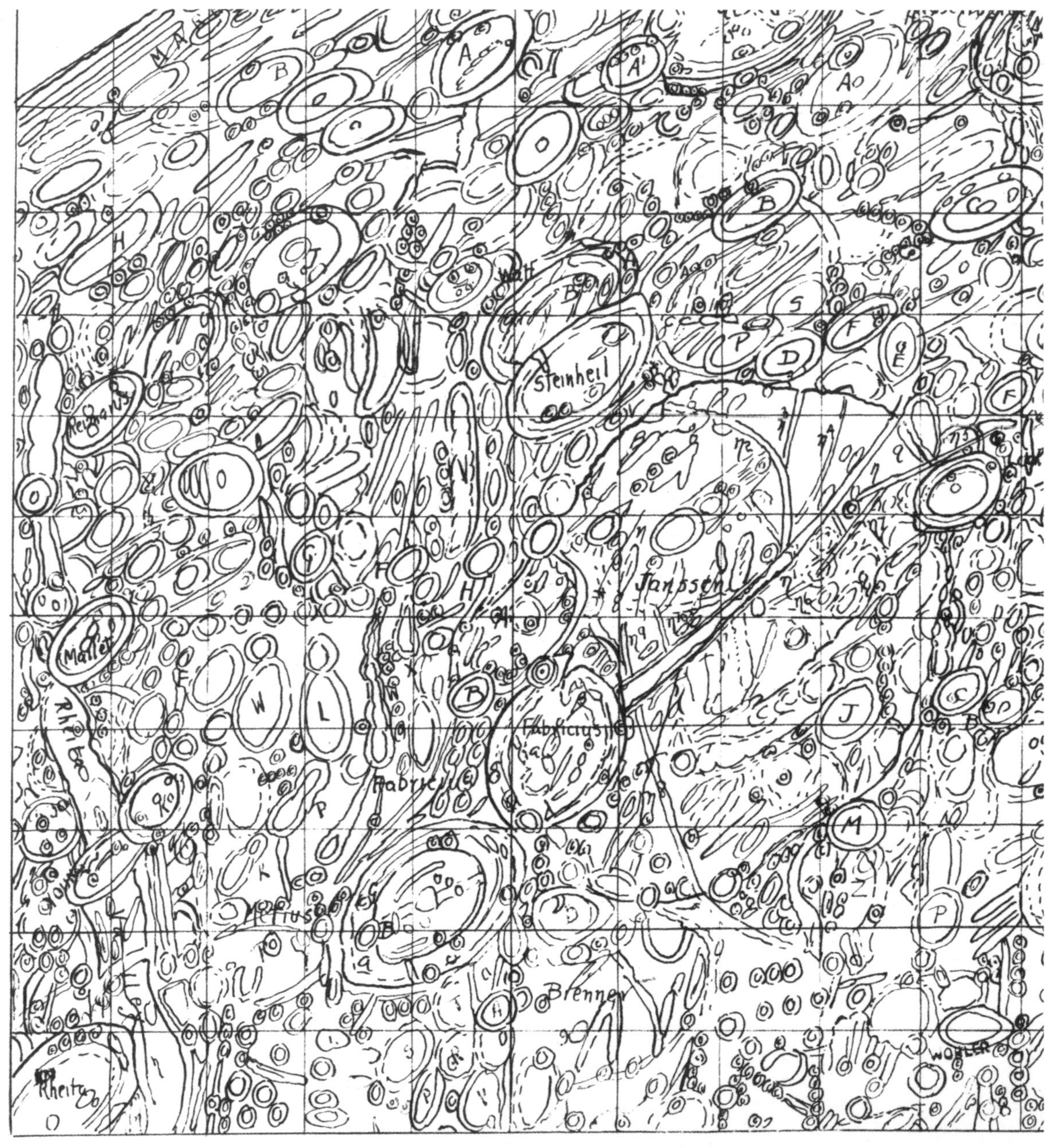
KEY:- Craterlet, Craterpit, Cratercone, Hillock, Clefts, Ridges, Light Streaks, Dark Variable Spots.
Steinheil
Janssen
Fabricius
Mallet
Brenner
100m Reproduction
THIRD EDITION

SECTION:
XXIV:IV

KEY:- ⊚ Craterlet, ⊙ Craterpit, ⊕ Cratercone, o Hillock, Clefts, ═ Ridges, ······ Light Streaks, # Dark Variable Spots.

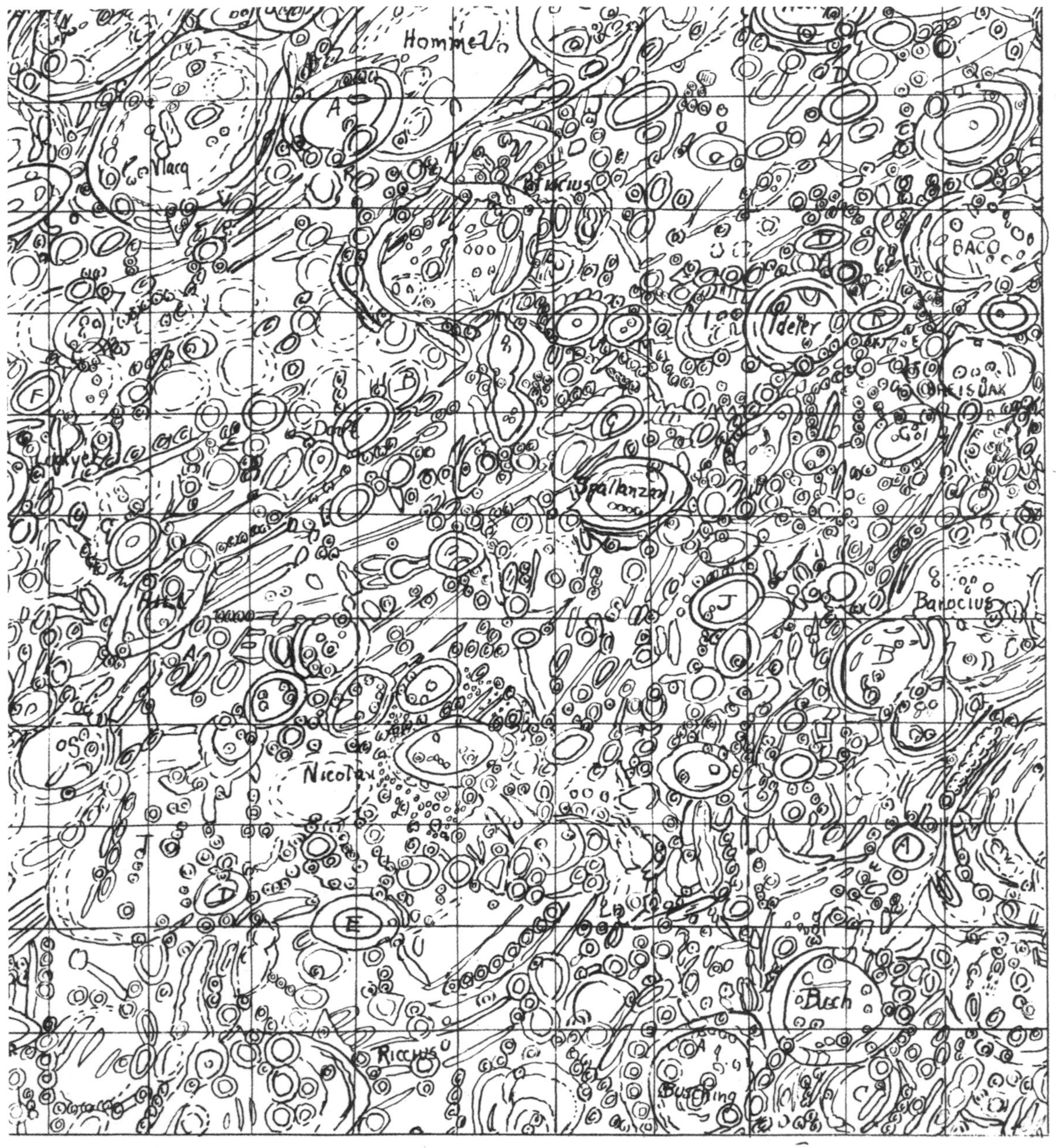

PRINTED BY W. F. STANLEY & Cº LTD 13, RAILWAY APPROACH, LONDON BRIDGE S. E. 1.

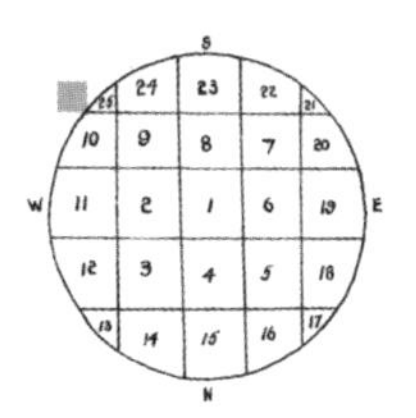

SECTION:
XXV:I

KEY:- ◎ Craterlet, ⊙ Craterpit, ⊕ Cratercone, ○ Hillock, Clefts, ═ Ridges, ····· Light Streaks, Dark Variable Spots.

8

100 inch Reproduction. THIRD EDITION.

Based on observations & measures by the A

KEY:- ◉ Craterlet, ⊙ Craterpit, ⊕ Cratercone, o Hillock, Clefts, ═ Ridges, ······ Light Streaks, Dark Variable Spots.

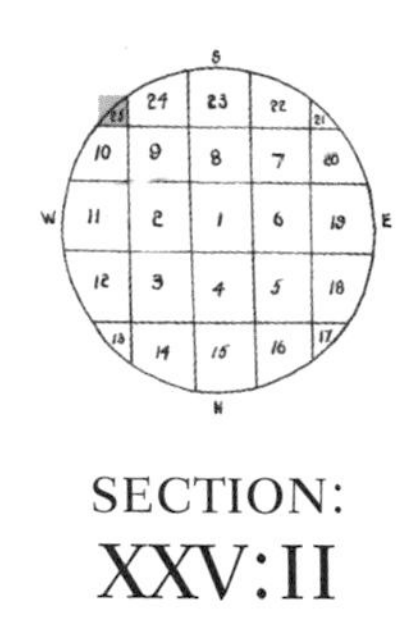

SECTION:
XXV:II

by the Author.

Drawn by H. Percy Wilkins. F.R.A.S.-1951.

PRINTED BY W.F. STANLEY & Co LTD 13 RAILWAY APPROACH, LONDON BRIDGE S.E.I.

At the very edge of the visible surface of the Moon, **Mare Australe** is one of the more difficult lunar mare to observe. It appears as an irregular dark patch, as it is marked by a number of later crater impacts.

Index to Formations in the 300-in. Lunar Map

Roman figures indicate Map Sections. LR = Special Chart of Libratory Regions. SC = Chart of Libratory Regions, drawn on Stereographic Projection. Third Edition.

Cayley	II
Celsius	IX
Censorinus	II
Cepheus	XIV
Chacornac	III
Challis	XV, LR
Chevallier	XIV
Chladni	I
Cichus	VII
Clairaut	XXIII
Clausius	VII-XXII
Clavius	XXIII
Cleomedes	XII
Cleostratus	XVI, LR
Colombo	X
Comas Solá	XX
Condamine	XVI
Condorcet	XII
Conon	IV
Cook	X
Cooke	XII
Copernicus	VI
Corderilla Mts.	XX
Cortés	LR, SC
Crisium, Mare	XI-XII
Crozier	X
Crüger	XX
Curtius	XXIII
Cusanus	XIV
Cuvier	XXIII
Cyrillus	IX
Cysatus	XXIII
Daguerre	II
D'Alembert Mts.	XIX, LR, SC
Damoiseau	XIX
Daniell	III
Darney	VII
D'Arrest	II
Darwin	XX
Da Vinci	XI
Davy	VIII
Dawes	III
De Bergerac	V
Debes	XII
Dechen	XVII
De Gasparis	XX
Delambre	II
De la Rue	XIV
Delaunay	VIII
Delisle	V
Delmotte	XII
Deluc	XXIII
Dembowsky	I
Democritus	XIV
Demonax	XXIII-XXIV, LR
De Morgan	II
Descartes	IX
Deseilligny	III
Deville, Pr.	XV
Dionysius	II
Diophantus	V
Di Vico	XX
Dollond	II
Donati	VIII
Döppelmayer	VII
Doerfel Mts.	XXII, SC, LR
Dove	XXIV
Draper	V
Drebbel	XXII
Drygalski	LR, SC
Dunthorne	VII
Dyson, Mount	V
Egede	XV
Eichstädt	XX, LR
Einmart	XII
Elger	VII
Emley	XXII
Encke	VI-XIX
Endymion	XIV
Epidemiarum, P.	VII
Epigenes	XV
Epimenides,	XXII
Eratosthenes	IV
Esquivel	LR, SC
Euclides	VI
Euctemon	XV
Eudoxus	XIV-XV
Euler	V
Fabricius	XXIV
Faraday	XXIII
Fauth	VI
Faye	VIII
Febrer	XI, LR, SC
Fermat	IX
Fernelius	XXIII
Feuillé	IV
Firminicus	XI
Flammarion	I
Flamsteed	XIX
Fœcunditatis, M.	X-XI
Fontana	XX
Fontenelle	XV
Foucault	XVI
Fourier	XX
Fracastorius	IX
Fra Mauro	VI
Franklin	XIV
Franz	XII
Fraunhöfer	XXV
Fresnel, Cape	IV
Frigoris, Mare	XIV-XV-XVI
Furnerius	X-XXV
Galileo	XIX
Galle	XIV
Galvini	XVII, LR, SC
Gambart	VI
Gärtner	XIV
Gassendi	VII-XX
Gaudibert	XI
Gauricus	VIII
Gay-Lussac	V
Gay-Lussac, Sinus	V
Gauss	XII-XIII, LR
Geber	IX
Geminus	XII
Gemma Frisius	VIII
Gérard	XVII, LR, SC
Gioja	XV, LR, SC
Glaisher	XII
Goclenius	XI
Godin	I
Goldschmidt	XV
Goodacre	IX
Gould	VII
Green	LR, SC
Grimaldi	XIX, LR
Grove	XIV
Gruemberger	XXIII
Gruithuisen	V
Guericke	VII
Guttemberg	XI
Gylden	I
Haas	XV
Hadley, Mount	IV
Hæmus Mts.	III
Hagecius,	XXIV
Hahn	XII
Haidinger	XXII
Hainzel	XXII
Hall	III
Halley	I
Hanno	XXIV, LR
Hallowes	XIII
Hansen	XI
Hansteen	XX
Harbinger Mts.	V
Harding	XVII
Harpalus	XVI
Hase	X
Hausen	XXII, LR, SC
Heinsius	XXII
Heis	V
Hekatæus	X
Helicon	XVI
Hell	VIII
Helmholtz	XXIV
Henry, Frères	XX
Heraclides, Pr.	XVI
Heraclitus	XXIII
Hercules	XIV
Hercynian Mts.	XVIII
Herigonius	VII
Hermann	XIX
Herodotus	XVIII
Herschel	I
Herschel, C.	V
Herschel, J.	XVI
Hesiodus	VII
Hevel	XIX
Hiemis, Mare	XIX, LR
Hind	I
Hippalus	VII
Hipparchus	I
Holden	X
Hommel	XXIV
Hooke	XIII

Hörbiger	VIII
Horrebow	XVI
Horrocks	I
Hortensius	VI
Huggins	XXIII
Humboldt, W.	X, SC
Humboldtianum, M.	XIV, LR, SC
Humorum, M.	VII-XX
Huygens	IV
Hyginus	I
Hypatia	II
Ibáñez	LR, SC
Ideler	XXIV
Imbrium, Mare	IV, V, XV, XVI
Incognito, Mare	SC
Inghirami	XXI, LR, SC
Iridum, Sinus	XVI
Isidorus	II
Jacobi	XXIII
Jansen	III
Janssen	XXIV
Jiyah	XX
Juan	LR, SC
Julius Cæsar	II
Jura Mountains	XVI
Kaiser	VIII
Kane	XV
Kästner	XI, LR, SC
Kelvin, Cape	VII
Kepler	VI-XIX
Kies	VII
Kinau	XXIII
Kirch	XV
Kircher	XXII, LR
Kirchhoff	III
Klaproth	XXIII
Klein	VIII
König	VII
Kraft	XVIII
Krieger	XVIII
Krosigk	V
Krusenstern	VIII
Kunowsky	VI
Lacaille	VIII
Lacroix	XXI
Lade	I
Lagalla	XXII
Lagrange	XX, LR
La Hire	V
Lalande	I
Lambert	V
Lamèch	XV
Lamont	II
Landsberg	VI
Langrenus	XI
La Paz	XX
Lapeyrouse	XI
Laplace, Pr.	XVI
Lassell	VIII
Lavinium, Pr.	XII
Lavoisier	XVII, LR, SC
Lee	VII

Legendre	X
Legentil	XXII, LR
Lehmann	XXI
Leibnitz Mts.	XXIII, LR, SC
Le Monnier	III
Lepaute	VII
Letronne	XIX, XX
Leverrier	XVI
Lexell	VIII
Licetus	XXIII
Lichtenberg	XVIII
Lick	XII
Liddiard	XII, LR, SC
Liebig	XX
Lilius	XXIII
Lindenau	IX
Linné	IV
Lippershey	VIII
Littrow	III
Löhrmann	XIX
Lohse	X
Longomontanus	XXII
Lockyer	XXIV
Louville	XVI
Lowe	XIX, LR, SC
Lower	XII
Loewy	VII
Lubbock	XI
Lubiniezky	VII
Luther	III
Lyell	XII
Maclure	X
MacDonald	XVIII
Maclaurin	XI
Maclear	II
Macrobius	XII
Mädler	II
Magelhæns	X
Maginus	XXIII
Main	XV, LR
Mairan	XVI
Malapert	XXIII, LR
Mallet	XXIV
Manilius	IV
Manners	II
Manzinus	XXIII
Maraldi	III
Marco Polo	IV
Marginis, M.	XII, LR, SC
Marinus	XXV, LR
Marius	XVIII
Marth	VII
Maskelyne	II
Mason	XIV
Maupertuis	XVI
Maurolycus	XXIII
Maury	XIV
Mayer, C.	XV
Mayer, T.	V
Medii, Sinus	I
Mee	XXII
Menelaus	III
Mercator	VII
Mercurius	XIII
Mersenius	XX
Messala	XIII
Messier	XI
Metius	XXIV

Meton	XV, LR
Milichius	VI
Millás	VIII
Miller	XXIII
Mitchell	XIV
Moigno	XV
Möltke	II
Monge	X
Montanari	XXII
Moore	VII
Moretus	XXIII
Mortis, Lacus	XIV
Mösting	I
Möstlin	XIX
Mouchez	XV
Müller	I
Murchison	I
Mutus	XXIV
Najerá	LR, SC
Nansen	LR, SC
Nasirredin	XXIII
Nasmyth	XXII
Naumann	XVIII
Neander	IX
Nearch	XXIV
Nebularum, P.	IV, XV
Nectaris, Mare	II, IX
Neison	XV
Neper	XI, XII, LR, SC
Neumayer	XXIV, LR, SC
Newcombe	XII
Newton	XXIII, LR
Nicolai	XXIV
Nicollet	VII
Nöggerath	XXII
Nonius	VIII
Novellas	VII
Novum, Mare	XII
Nubium, Mare	VI, VII, VIII
Œnopides	XVI
Œrsted	XIV
O'Kell	XIV
Oken	XXV, LR, SC
Olbers	XIX, LR
Olivium, Pr.	XII
Opelt	VII
Oppolzer	I
Oriani	XII
Orientalis, M.	XX, LR, SC
Orontius	XXIII
Palisa	I
Palitzsch	X
Pallas	I
Palmieri	XX
Paluzíe	XII, LR, SC
Parrot	VIII
Parry	VI
Parvum, Mare	XXI, LR, SC
Peary	LR, SC
Peirce	XII
Peirescius	XXV
Pentland	XXIII
Percy Mts.	XX
Petavius	X
Petermann	XIV
Phillips	X
Philolaus	XV, LR
Phocylides	XXII
Piazzi	XXI

APPENDIX

Hugh Percy Wilkins's selenographical observations

After the publication of this edition of his map, Hugh Percy Wilkins continued to observe the Moon. Observation notebooks, now in the collection of the Museum, show the detailed sketches of individual features which he continued to make, alongside his notes on observing conditions. Some of these show how the changing appearance of shadows over time helped to reveal the shape of different features. A selection of his sketches are reproduced here.

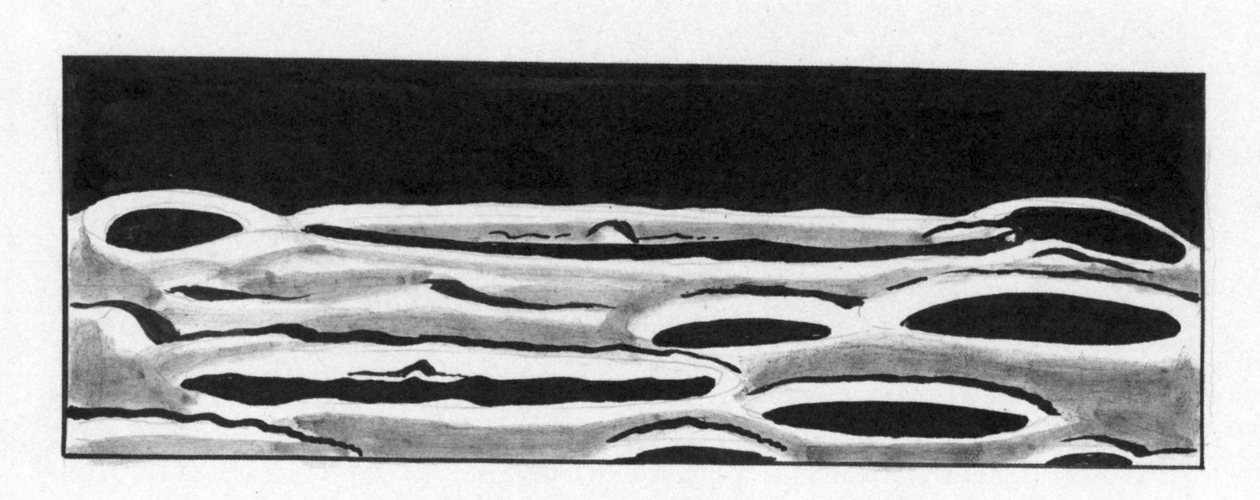

Drygalski & Region.
Aug. 3. 1955. 15¼ in. x 250.
Libration not very favourable but the central hill well seen. The walls seemed fairly regular. On the north a crater abuts & either a spur runs on to the floor or there is a small crater. On the south is a large crater.

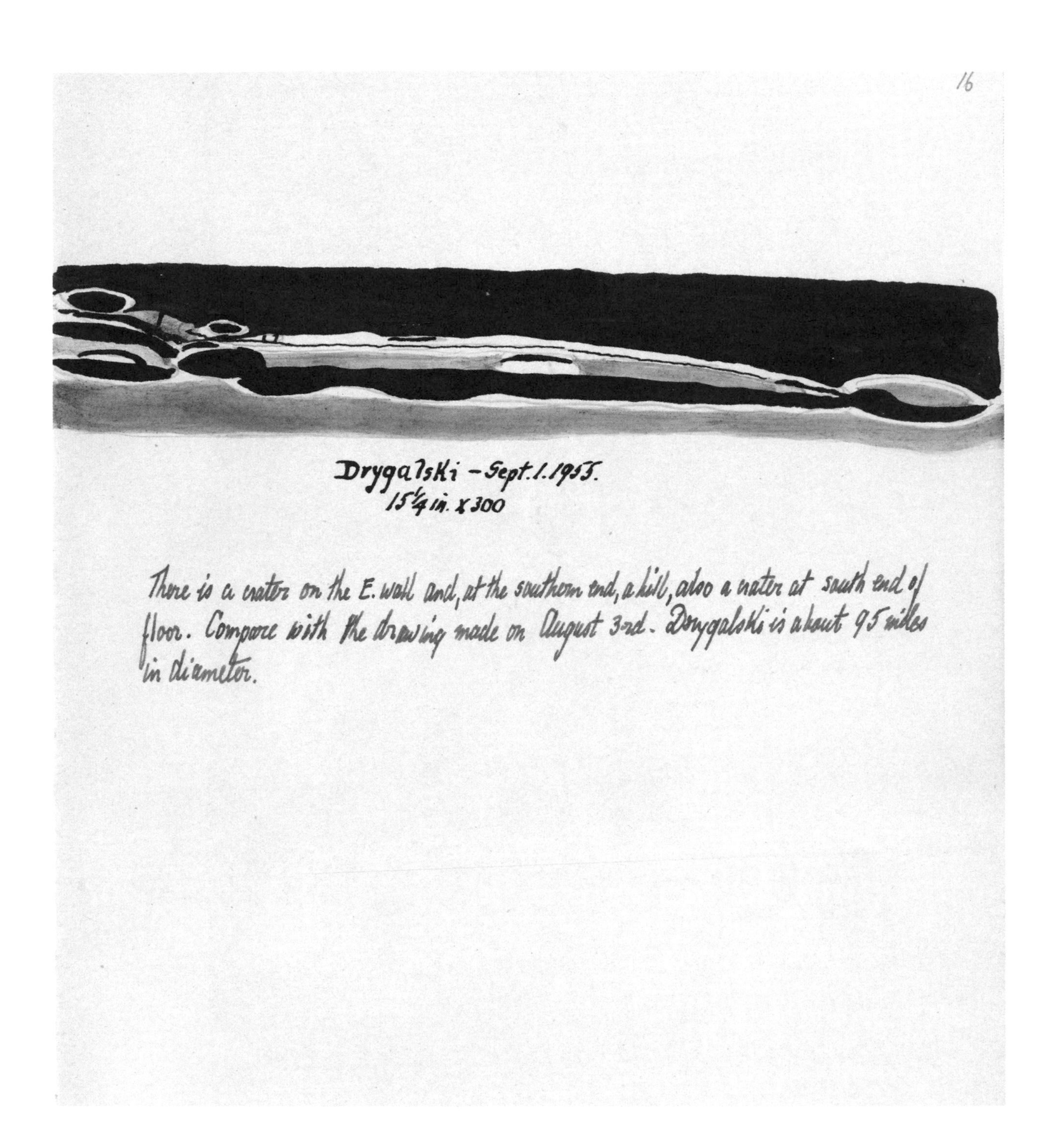
16
Drygalski – Sept. 1. 1955.
15¼ in. x 300
There is a crater on the E. wall and, at the southern end, a hill, also a crater at south end of floor. Compare with the drawing made on August 3rd. Drygalski is about 95 miles in diameter.

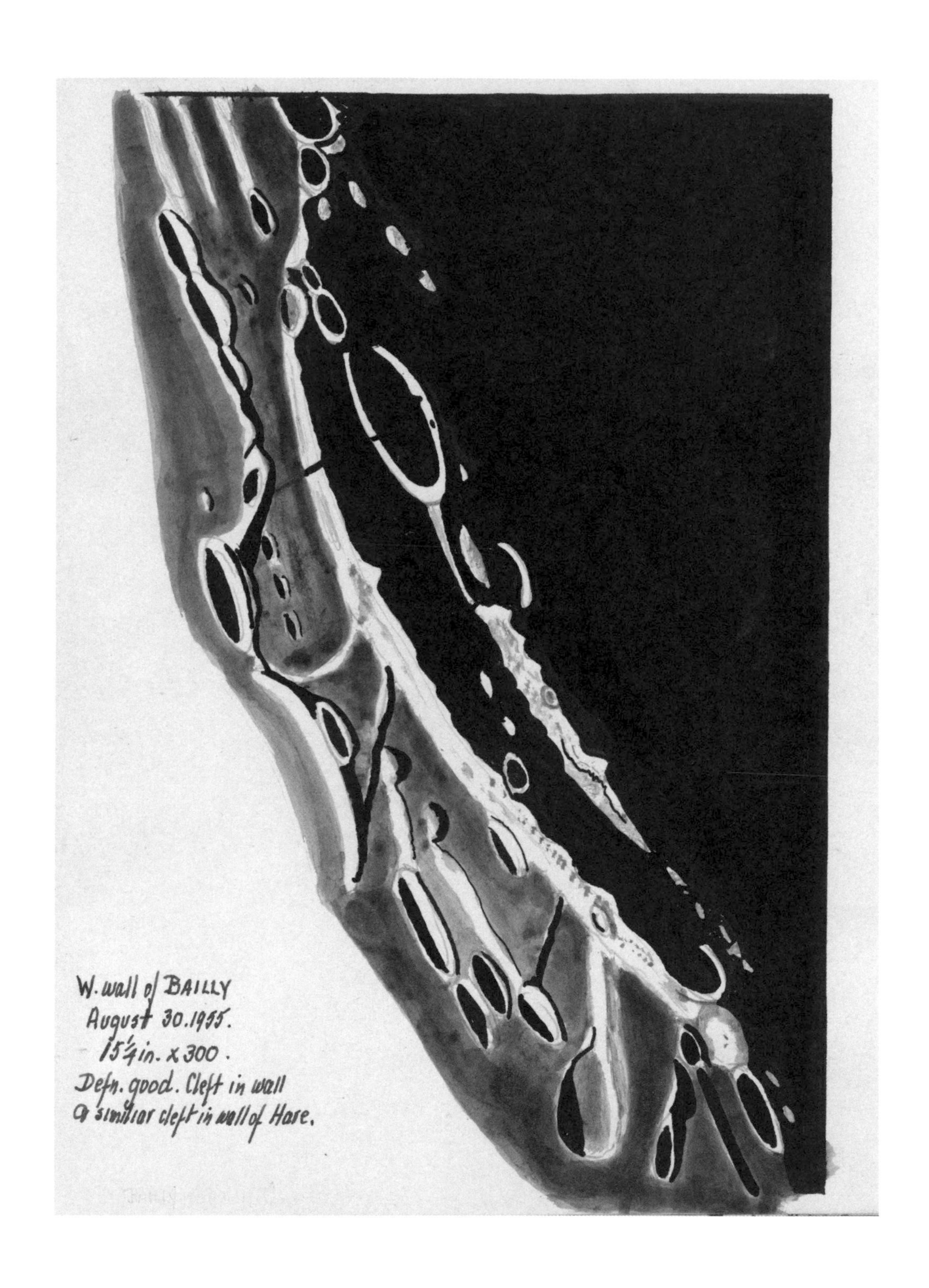
W. wall of BAILLY
August 30.1955.
15¼ in. x 300.
Defn. good. Cleft in wall
a similiar cleft in wall of Hase.

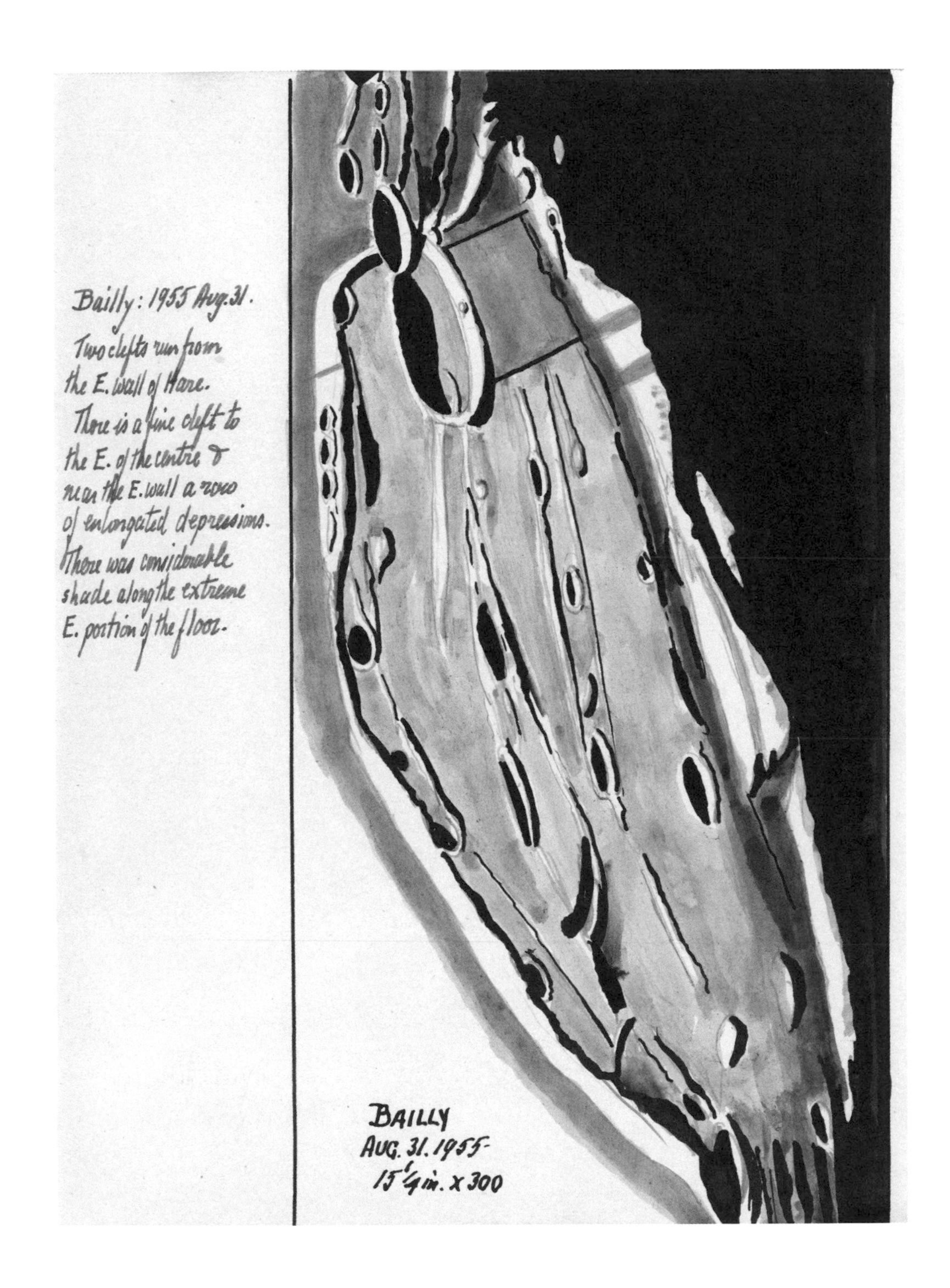
Bailly: 1955 Aug. 31.
Two clefts run from the E. wall of Hare.
There is a fine cleft to the E. of the centre & near the E. wall a row of elongated depressions.
There was considerable shade along the extreme E. portion of the floor.
BAILLY
AUG. 31. 1955
15¼ in. x 300

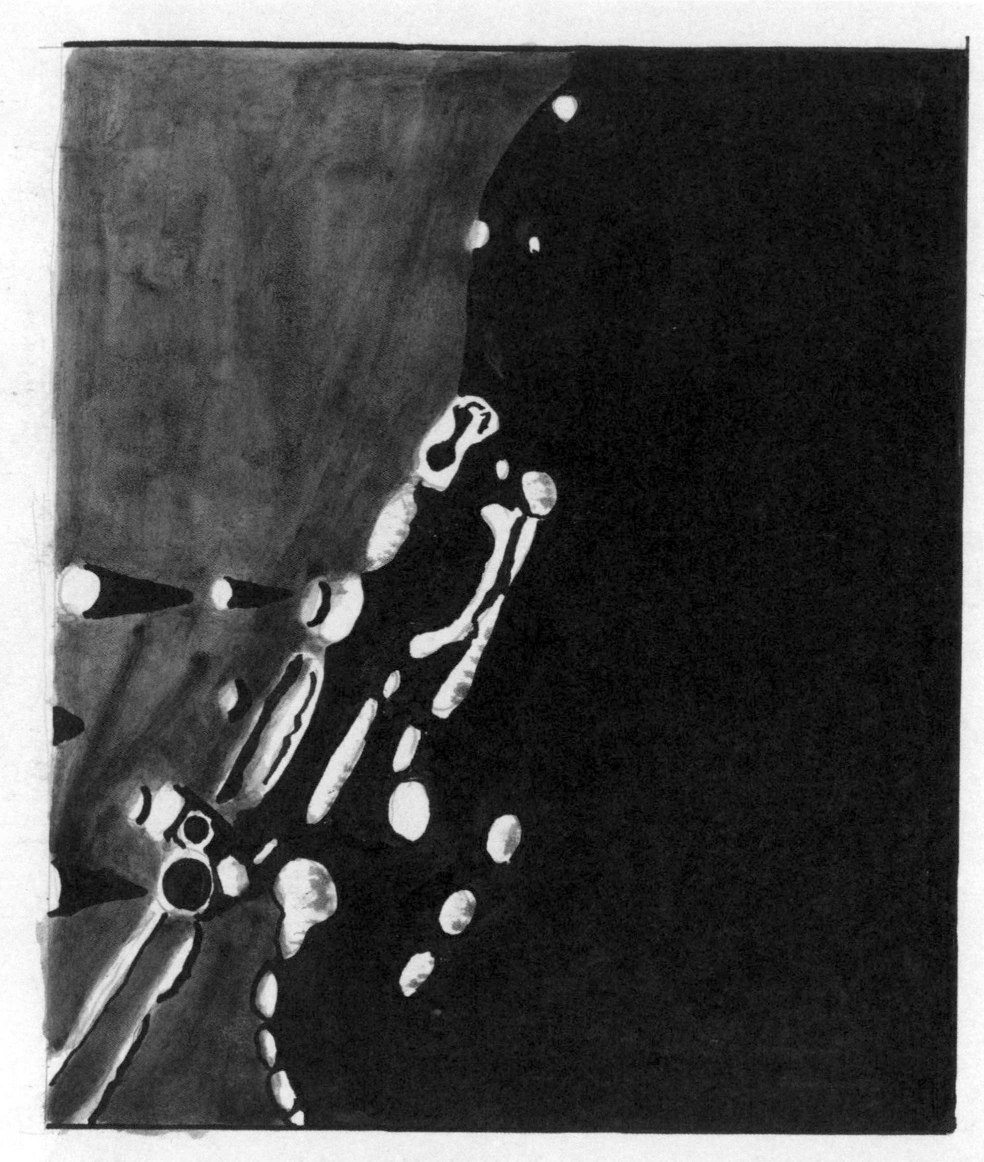

Cape Laplace. August. 27. 1955. 8h15m. U.T. 3½in 'Ross' Refractor.
Moon low, definition moderate, surface very dark, terminator on the south sharply curved. The most southerly hill very brilliant.

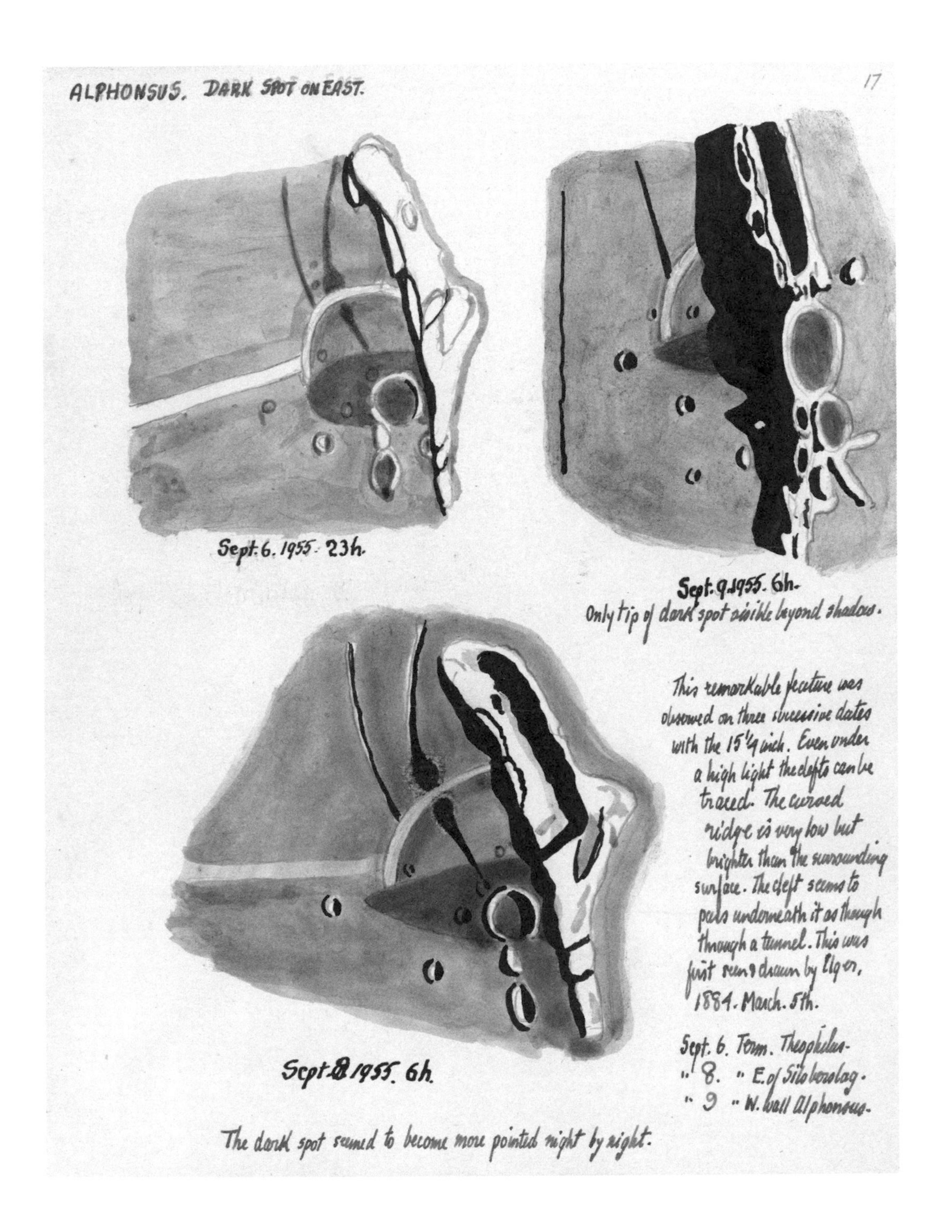
ALPHONSUS. DARK SPOT ON EAST.
17
Sept. 6. 1955. 23h.
Sept. 9. 1955. 6h.
Only tip of dark spot visible beyond shadow.
This remarkable feature was observed on three successive dates with the 15¼ inch. Even under a high light the clefts can be traced. The curved ridge is very low but brighter than the surrounding surface. The cleft seems to pass underneath it as though through a tunnel. This was first seen & drawn by Elger, 1884. March. 5th.
Sept. 6. Term. Theophilus.
" 8. " E. of Sitobozlag.
" 9 " W. wall Alphonsus.
Sept. 8 1955. 6h.
The dark spot seemed to become more pointed night by night.

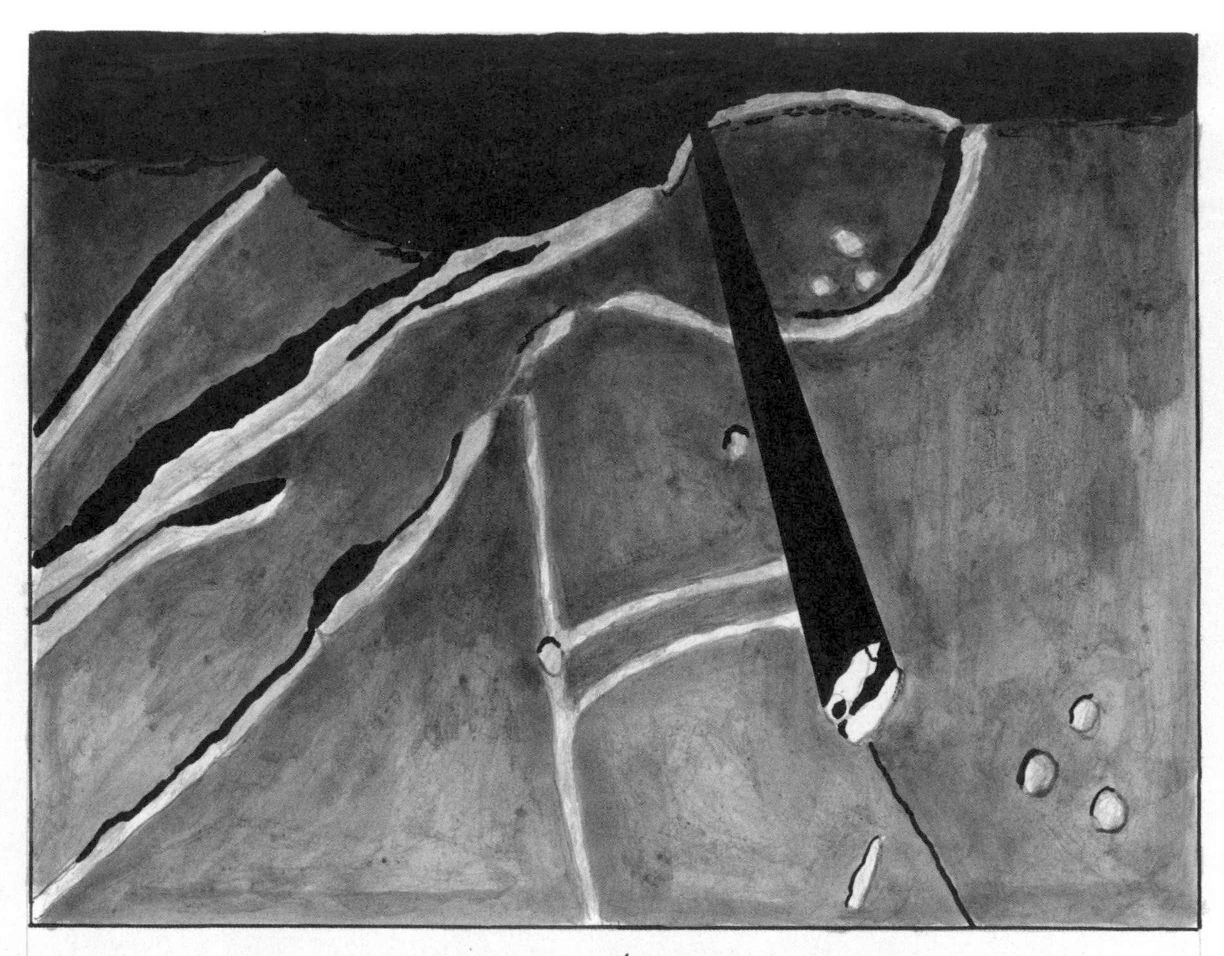

LA HIRE. SEPT. 11. 1955. 6h. S.T. 15¼in. x280.
The mountain was very brilliant this morning & cast a long shadow to the ridges on terminator.
There is a depression on the N. slope of La Hire, & a cleft? running N.E.

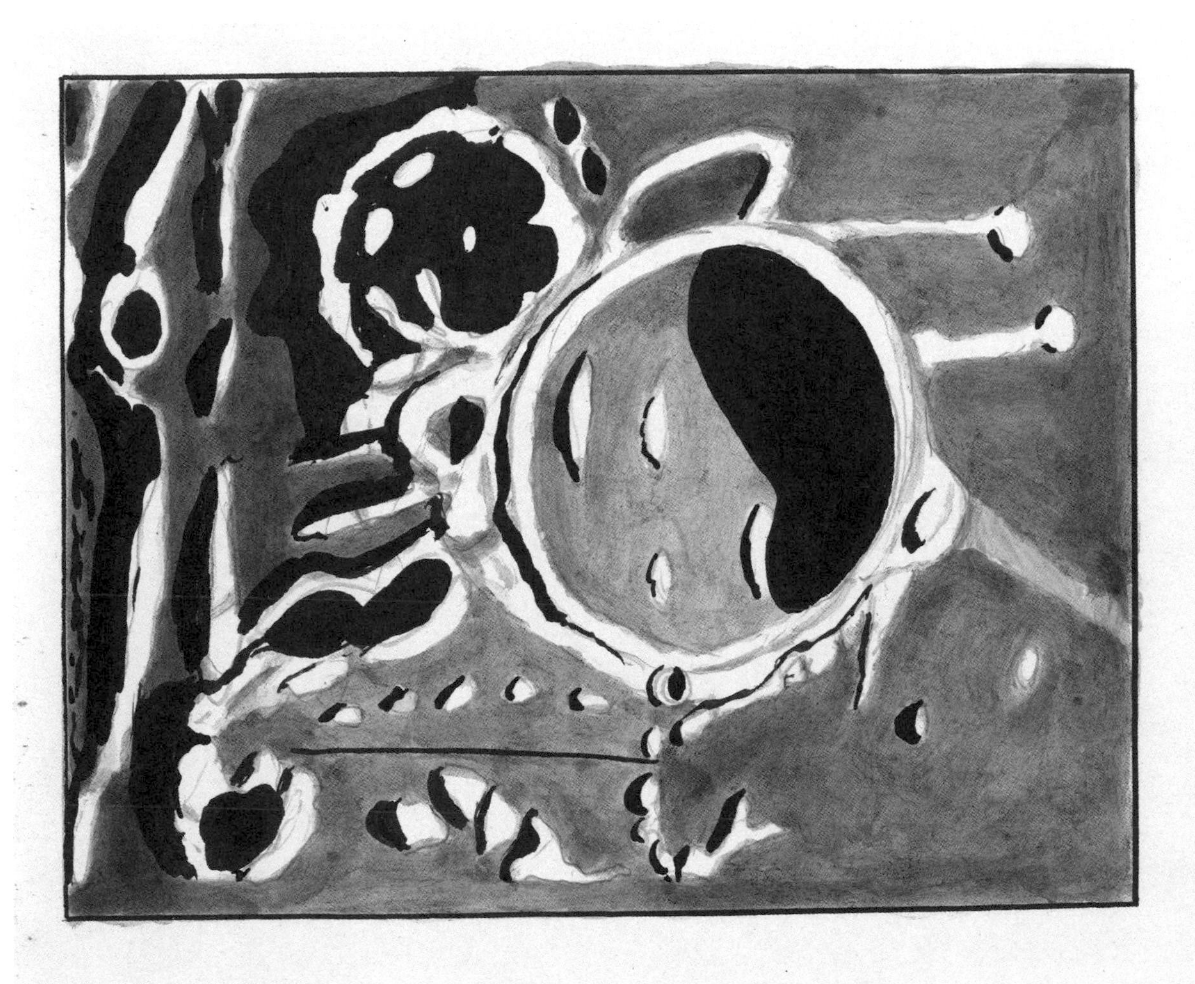

Tobias Mayer. Sept. 11. 1955. 6h. 15¼in. x 280.

The interior shadow, under evening illumination, is peculiar. Crater A, on the S.W. is deep & shadow filled. A craterlet on the N. crest of Tobias Mayer & on its N. a cleft runs E. and W. Definition variable, good at times. Some delicate details on the N.W. omitted owing to cloud & dawn coming on.

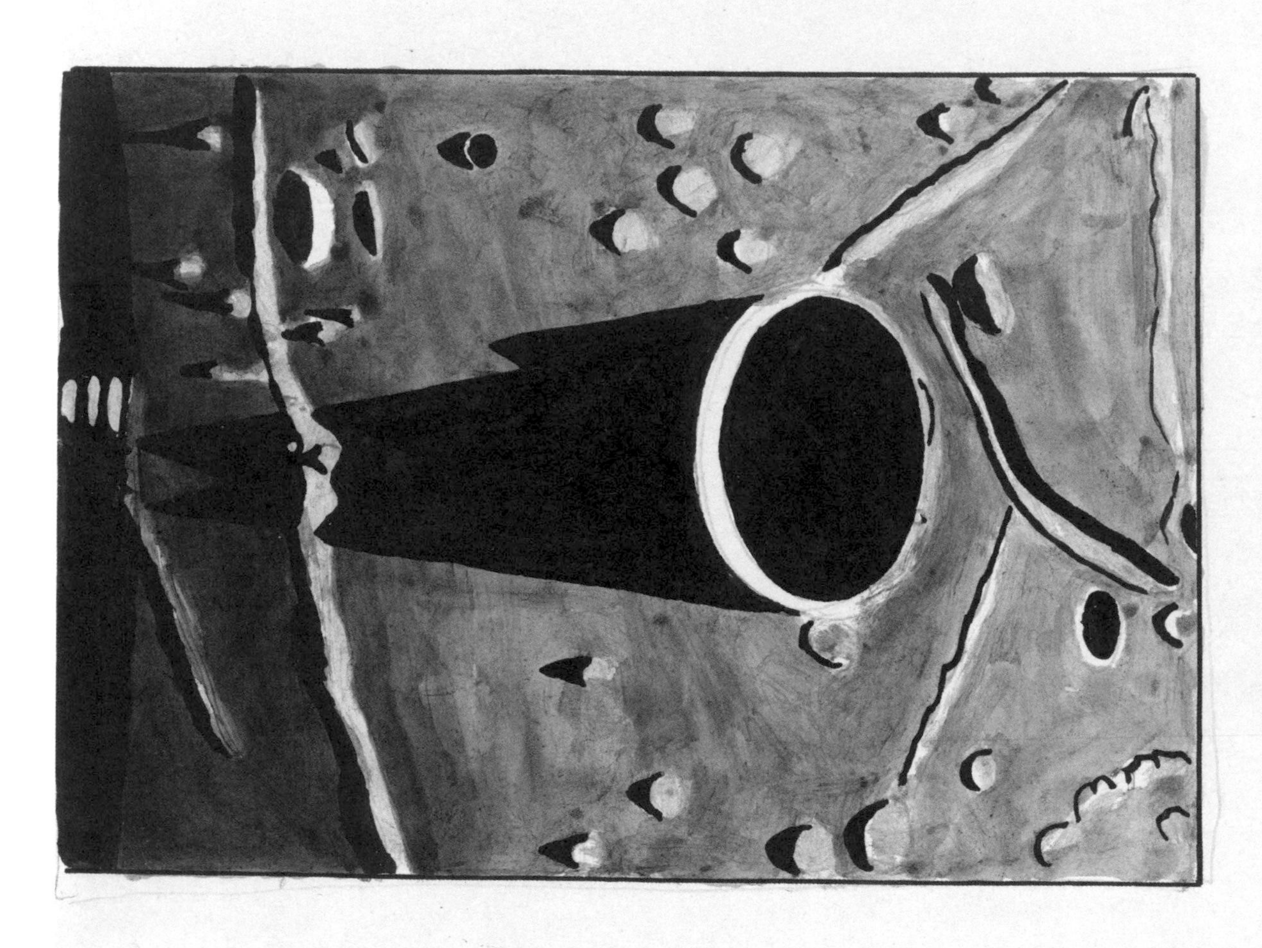

KEPLER. SEPT. 12. 1955. 5h.
15¼ in. × 280.

At first sight the shadow seemed to reach the terminator, but it was then seen that the Kepler shadow ended at a ridge & this ridge cast its own shadow which happened to be a prolongation of that of Kepler.

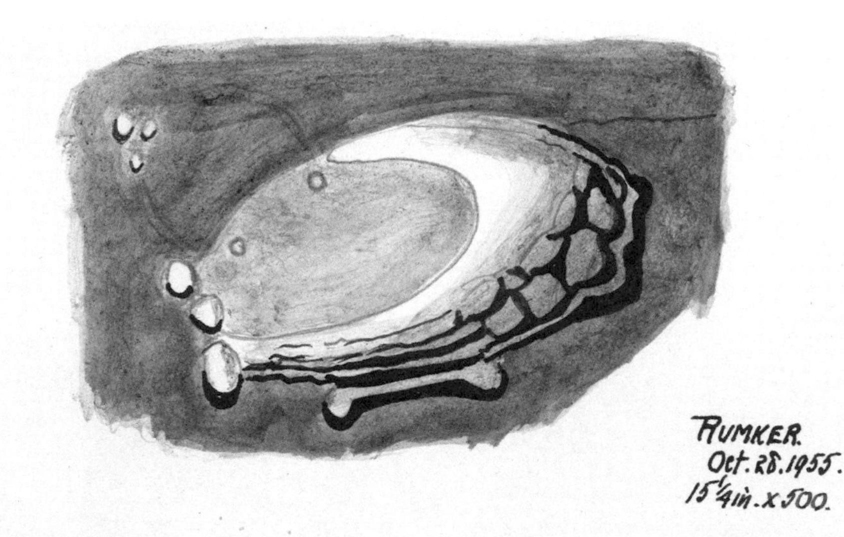

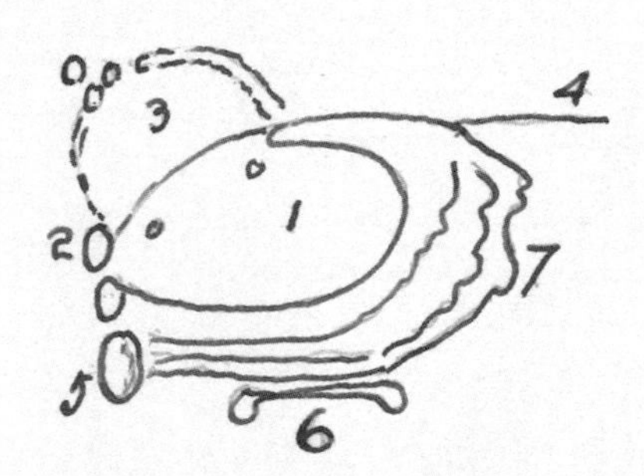

The best view I have had of this peculiar & fascinating object. Definition good using the American eyepiece & Barlow. It resembles a sloping ring with an interior 'lagoon'-1. The hill 2, was very brilliant. 3 is an old ring & 4 a cleft-like marking previously seen.
The largest rocky mass is 5 & 6 is a detached elevation of peculiar shape. Rumker is highest on the east where is a line of dark coloured cliffs & rocks-7.

The 'lagoon'-1, had a faint bluish tint; two objects were seen on it. From the high ground on the east the formation slopes down towards the 'lagoon.'

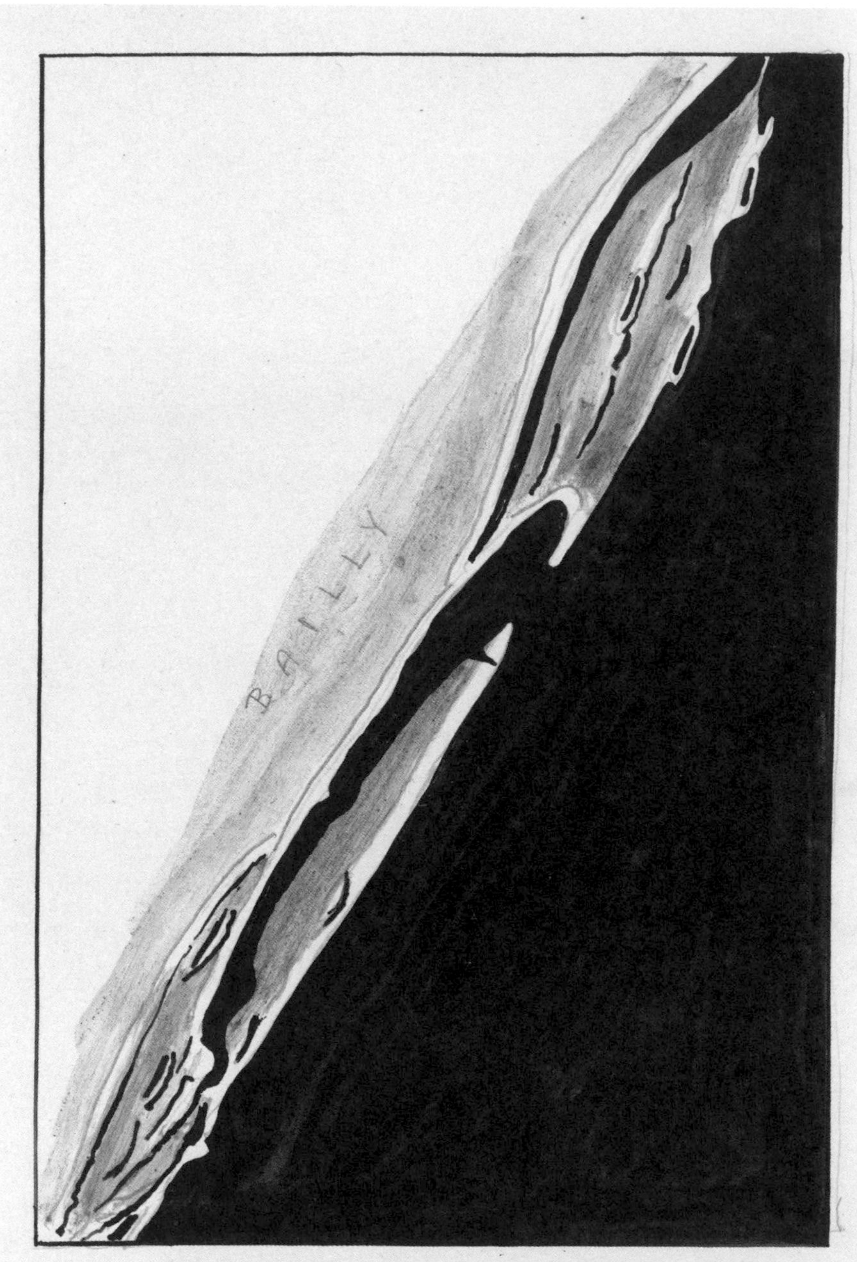

Rings directly east of Bailly. (S. end opposite N. end Bailly)
Oct. 30. 1955. 20h. $15\frac{1}{4}$ in. x 500.

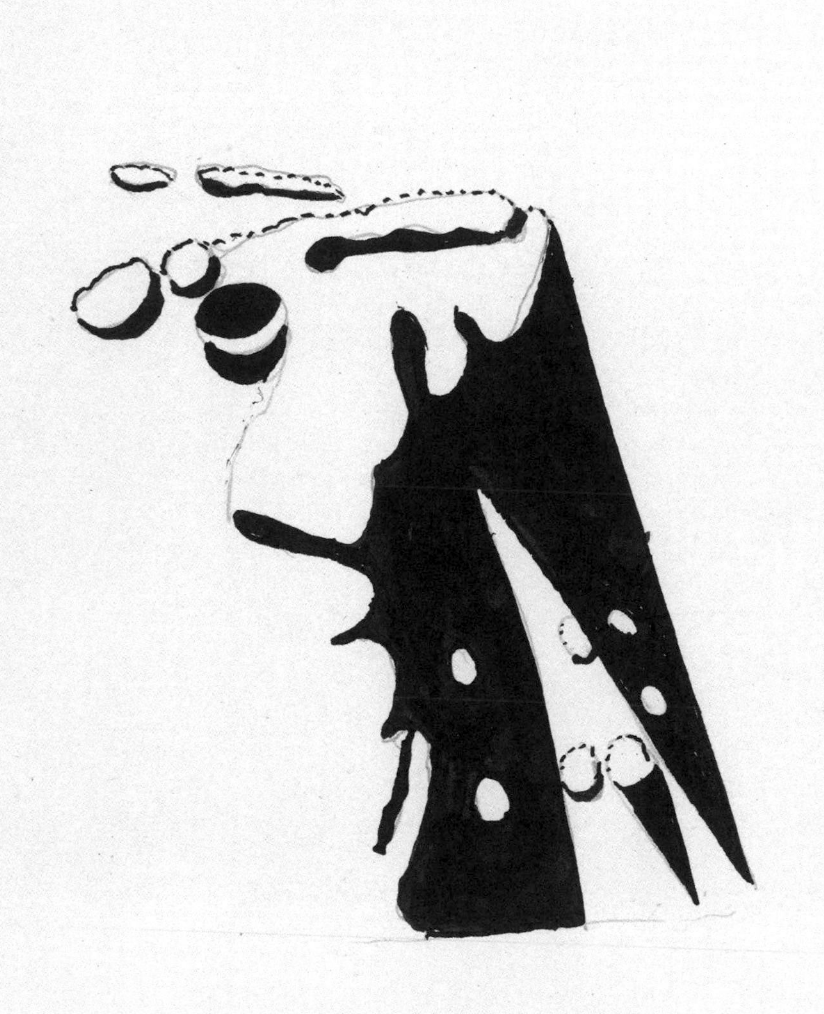

C. Laplace. Nov. 24. 1956. 7.30p.m
$15\frac{1}{4}$ in. x 500.

First published in 2019 by the National Maritime Museum,
Park Row, Greenwich, London SE10 9NF

ISBN: 9781906367602

At the heart of the UNESCO World Heritage Site of Maritime Greenwich are the four world-class attractions of Royal Museums Greenwich – the National Maritime Museum, the Royal Observatory, the Queen's House and *Cutty Sark*.

www.rmg.co.uk

A CIP catalogue record for this book is available from the British Library.

Designed by carrdesignstudio.com
Printed and bound in the UK by Gomer Press

Created from a 100-inch reproduction in 25 sheets of Hugh Percy Wilkins's 300-inch *Map of the Moon*, with title page and index, 1951 (object ID: ZBA4573)

Sketches in appendix taken from 'Selenographical observations by H.P. Wilkins, FRAS' (object ID: REC/69/7/3)

The Royal Museums Greenwich Picture Library contains over 45,000 images and photographs depicting notable ships, trade and empire, astronomy, exploration, navigation and time. The images within this book are available to license, please contact pictures@rmg.co.uk for more information.